No-Frills Statistics

No-Frills Statistics

A Guide for the First-Year Student

SUSAN H. GRAY

A Littlefield, Adams
Quality Paperback

Rowman & Littlefield Publishers, Inc.

Published as A HELIX BOOK
in the United States of America in 1984
by Rowman & Allanheld
(A division of Littlefield, Adams & Company)

Library of Congress Cataloging in Publication Data

Gray, Susan H.
No-frills statistics.

1. Statistics. I. Title.
QA276.12.G7 1983 519.5 83–11015
ISBN 0–8226–0380–2

Reprinted in 1989 by Rowman & Littlefield Publishers, Inc.,
as a Littlefield Adams Quality Paperback

Printed in the United States of America

Table of Contents

List of Tables

Introduction

You may be studying statistics for several reasons: Is it a course requirement for your major or a course you will need to have taken in order to enter graduate school? Many reasons are not entirely voluntary and you may be viewing the study of statistics with a combination of fear and loathing. This fear and loathing is often only intensified when you try to make sense of the statistical formulas that you encounter. Math creates anxiety in a lot of students and this anxiety often gets in the way of your studies. This manual attempts to allay your anxiety, fear, and loathing.

Statistics can be fun if you can see the practical applications of what you learn. This manual exposes you to common statistical procedures with step-by-step illustrations of their practical application to a problem from either the social, behavioral or educational sciences. Unlike most other manuals, it is not necessary for you to learn or remember any algebra. Symbols and formulas are not introduced until the end of the instructions for a particular procedure. You can then look at the formula which represents what you have already learned how to do. But if the symbols in the formula make you anxious, you can instead use the instructions, which explain how to use the formula in calculations, and ignore the formula. Each chapter also contains review questions and review exercises, which provide illustrative solutions to problems.

This manual is called "No-Frills Statistics" because the instructions provided for carrying out statistical techniques are very basic. The derivations of formulas are not reviewed and generally the most common and/or the easiest computing formula is used for each technique.

This manual reviews the most basic techniques. Enough basics are provided to prepare undergraduates in a one-semester statistics course to perform well. It also provides enough statistical background for more advanced learning from more traditional texts. Most chapters can be used independently of all other chapters and you can re-order chapters according to your statistics syllabus.

The content of this manual is practical rather than philosophical. It emphasizes "how to" rather than why. It will be best used as a study outline when you are preparing homework assignments and are preparing for exams. It is also useful as a review tool for those of you who must re-learn statistical techniques that have been learned and forgotten.

I am grateful to the literary executor of the late Sir Ronald A. Fisher, F.R.S., to Dr. Frank Yates, F.R.S., and to Longman Group, Ltd., London for permission to adapt and reprint Tables III, IV, and V from their book *Statistical Tables for Biological, Agricultural and Medical Research* (Sixth Edition, 1974). I am also grateful to A. Hald and to John Wiley & Sons, Inc., New York, for permission to adapt and reprint Table I from the book *Statistical Tables and Formulas* (1952).

1

Frequency Distributions

How to Arrange Information to Make it More Readable

Ungrouped Frequency Distribution

Let us suppose the following numerical values are scores on an examination:

81	87	89	92
86	83	88	90
93	89	86	87
90	89	85	88
84	85	87	84
86	87	89	92
90	88	85	87
90	87	89	88
91	86	88	86

This is called *raw data.* It is a simple listing of scores. There are 36 scores in this list. Most of the scores occur more than once, making this list somewhat difficult to read. It is also difficult to read because the scores are not arranged in order from lowest to highest (or from highest to lowest). The scores need to be organized.

The scores range from 81 to 93. This is a relatively small range of numbers. Generally, if you have a range of 15 scores or less, an ungrouped frequency distribution is the best way of organizing the information from the list of numerical values. In an ungrouped frequency distribution, the number of times each score occurs is indicated in a separate column (*a frequency column*).

To make an ungrouped frequency distribution, do the following:

1) Make a column in which you list all scores in the range from lowest to highest, with the lowest score at the bottom. Label this column "score" or "X". In this example, it should look like this:

X
93
92
91
90
89
88
87
86
85
84
83
82
81

2) Make an "f" (frequency) column next to the scores and indicate the number of times each score occurs in the raw data. If the score does not occur in the raw data, place a zero in the frequency column next to that score. Remember, the numbers in the frequency column should *sum* to the total number of scores in the distribution. Your ungrouped frequency distribution should look like this:

X	f
93	1
92	2
91	1
90	4
89	5
88	5
87	6
86	5
85	3
84	2
83	1
82	0
81	1
	N = 36

Grouped Frequency Distribution

Let us suppose that we have a much larger range of scores between the lowest and highest numerical values in the distribution. For example, the following are a group of scores on a nation-wide examination:

793	598	541	491	459	421	352
762	598	537	489	455	419	349
749	586	536	488	453	418	346
724	580	533	482	449	414	341
703	573	531	482	448	411	332
690	572	523	479	441	403	321
671	572	522	479	439	397	312
655	569	514	479	438	394	311
650	564	513	477	438	377	291
649	559	511	473	434	372	288
648	547	507	471	434	368	279
634	546	502	470	433	358	259
622	544	499	463	422	356	244
617	543	492	463	421	355	207

There are 98 scores in this list. Because of the large number of scores, the distribution is very difficult to read, even more difficult to read than the previous list of raw data in this chapter. The large range of scores from the lowest to the highest also makes any attempt to get a sense of the nature of the scores unwieldy. These scores are ordered from highest to lowest. If they were listed in just any order, the distribution of scores would be even more difficult to read. Because the lowest score is 207 and the highest score is 793, an ungrouped frequency distribution would require a list of almost 600 potential scores. This would also be difficult to read. These scores need to be organized differently.

The most common way of organizing such data to make it more readable is to construct a *grouped frequency distribution.* A grouped frequency distribution arranges scores in order, from lowest to highest. It then arranges these scores into *groups* of several scores each (called *class intervals*). A grouped frequency distribution shows the number of times a score occurs within each interval.

There are some general guidelines for constructing grouped frequency distributions. However, most important is to do what makes sense to make your distribution readable. In general, to construct a grouped frequency distribution, do the following:

1) Look for the lowest score and the highest score in the list of scores. In the list on p. 3, the lowest score is 207 and the highest score is 793.

2) Subtract the lowest score from the highest score and add one to find the range of scores. In this example, $793 - 207 = 586 + 1 = 587$.

3) You want to construct between 8 and 15 class intervals, usually. Try dividing the range of scores (in this case 587) by various numbers between 8 and 15 until you get a number equal to or fairly close to, but less than, a round number. 5, 10, 25, 50, 100 are each appropriate round numbers for different distributions. The number 2 is also often used. In this case, $587/15 = 39$; $587/14 = 42$; $587/13 = 45$; $587/12 = 49$ which approaches 50; $587/11 = 53$ which is greater than 50. Since you want a number equal to, or close to, but less than, a round number (50 in this case), 49 is the closest you can get.

4) When you get that round number, use the number between 8 and 15, that you used in the denominator when dividing in step 3, as the number of class intervals to be constructed. In this case, the number of class intervals is 12.

5) Use the round number you obtained or approached when dividing as the number of potential scores to be grouped together in each interval. In this case, 587/12 = 49, which can be rounded up to 50 potential scores for each class interval.

6) Begin your class intervals with a score equal to or less than, but close to, the lowest number in your list of scores. For example, in this case, we should begin with a number less than, but close to, 207, the lowest actual score in the distribution. Your class intervals should contain the number of potential scores per interval you have decided upon and should end with the last potential score for that interval. In this case, each of our class intervals should contain 50 potential scores. Your class interval should either begin or end with a round number, if you think this will make your grouped frequency tables more readable. In this case, we might decide to end on a round number and begin with the class interval 201–250. If we wanted to begin with a round number, our first class interval would be 200–249. If round numbers are not appropriate, use any other numbers that make sense. In this case, since the first class interval must include the lowest score which is 207, and since we are using class intervals with 50 potential scores in each, we will begin with 201–250, ending on a round number.

7) Go to the next number which is 251 and begin your next class interval. Again, end your class interval with the last potential score for that interval. In this case, the second class interval is 251–300, again including 50 potential scores.

8) Continue this procedure until you have included in your class intervals all the scores from the list. Make sure that all your intervals are of equal *width*. In this case, our twelfth and last class interval is 751–800, which includes the highest score of 793.

9) Arrange your class intervals from lowest to highest, with the lowest interval at the bottom. At the top of the column, label it "Interval," or "Class Interval," or "score."

10) Make another column next to the class intervals. Label this "f," which is the symbol for frequency or number of times

any of the scores within a particular interval has occurred in the list of scores (see the distribution on p. 3).

11) Count the number of scores from the list which fall within each class interval and put that number under the "f" (frequency) column next to each interval. If no scores occurred in a class interval, put a zero next to the interval. (Intervals only delineate *potential* scores. *The number of actual scores is in the frequency column* and there may not be any for a particular interval.) The numbers in the frequency column should sum to the total number of scores in the distribution. Keep in mind that the sum of your frequency column (represented by the symbol N) is the total number of scores, *not* the total number of class intervals.

Your grouped frequency distribution should look like this:

Class Interval	f
751–800	2
701–750	3
651–700	3
601–650	6
551–600	10
501–550	16
451–500	19
401–450	17
351–400	9
301–350	7
251–300	4
201–250	2
	N = 98

Cumulative Frequency Distribution

In subsequent chapters you will see the uses of the *cumulative frequency distribution*. It shows the number of times a score occurs *at or below* a given value. In other words, frequencies accumulate up the cumulative frequency column of the distribution. To make a cumulative frequency distribution do the following:

1) Create a third column in your frequency distribution labeled "cumulative f" (cumulative frequency).

2) In the first (lowest) row of this column, place the frequency corresponding to the lowest class interval. In the case of the above grouped frequency distribution, that frequency is 2.

3) In the second row, add the frequency of the second interval to the frequency of the first interval and place the sum in the cumulative f column. In this example, $2 + 4 = 6$.

4) Continue adding in each successive frequency to the sum as you continue up the column. In this case, in the third class interval, $6 + 7 = 13$; in the fourth class interval, $13 + 9 = 22$, etc.

5) The cumulative frequency of the last (highest) class interval should equal the total number of scores (that is, the sum of the f column). In this example, 98 is the cumulative frequency of the last class interval and is also the sum of the f column. A cumulative frequency distribution for the grouped frequency distribution in this chapter should look like this:

Class Interval	f	cumulative f
751–800	2	98
701–750	3	96
651–700	3	93
601–650	6	90
551–600	10	84
501–550	16	74
451–500	19	58
401–450	17	39
351–400	9	22
301–350	7	13
251–300	4	6
201–250	2	2
	N = 98	

A cumulative frequency distribution can also be constructed from an ungrouped frequency distribution. The procedure would be the same as with a grouped frequency distribution, only you would have scores, rather than class intervals, in the first column.

Review Questions

1. What is the difference between raw data and an ungrouped frequency distribution?
2. What is the difference between a grouped and ungrouped frequency distribution?
3. How do you decide how many class intervals to include in a grouped frequency distribution?
4. What do you do if no scores fall into one of the class intervals in your frequency distribution?
5. What does a cumulative frequency distribution do?
6. What should the cumulative frequency distribution of the highest interval or score equal?

Review Exercises

Exercise 1. In a survey of the number of television sets owned by a sample of American families, the following figures were obtained:

4	3	2	2	2
2	5	3	3	5
6	2	1	4	1
3	1	5	2	3
3	3	2	1	1
2	2	3	3	4
4	0	2	1	2
1	4	4	4	1
2	3	1	1	3
3	3	3	3	2

Construct a frequency distribution.

1) The numbers here range from 0 to 6. With this small a range, intervals would be awkward. Therefore, an ungrouped frequency distribution is more appropriate. A column in which

numbers in the range are listed from lowest to highest, with the lowest number at the bottom would look like this:

X
6
5
4
3
2
1
0

Do not forget to label your column "X" or "score."

2) In the "f" column next to the scores, we indicate the number of times each score occurs. We obtain the following:

X	f
6	1
5	3
4	7
3	15
2	13
1	10
0	1
	N = 50

Exercise 2. The following are the results of a survey of restaurant dining patterns among 95 New York City single people. Participants in the survey were asked to indicate the number of evenings in which they dined out over a three-week period. The responses were as follows:

3	6	3	3	3
11	7	7	6	7
4	10	11	10	8
21	5	6	6	1
8	18	9	3	13

6	4	5	4	6
4	15	14	5	5
0	9	7	7	7
5	5	5	9	4
10	7	7	4	7
7	2	9	13	11
4	2	5	15	8
3	6	10	8	2
2	9	6	7	7
6	5	3	2	1
8	12	6	5	5
5	6	2	1	3
12	4	7	6	7
7	3	3	7	10

Construct a frequency distribution.

1) The lowest score is 0. The highest score is 21.

2) The lowest score subtracted from the highest score plus one is $21 - 0 + 1 = 22$. This tells us the range of scores. There is a large enough range of scores here to make a grouped frequency distribution desirable.

3) $22/15 = 1.5$; $22/14 = 1.6$; $22/13 = 1.7$; $22/12 = 1.8$; $22/11 = 2.0$. When dividing the range of scores (in this case 22) by various numbers between 8 and 15, $22/11 =$ the whole number of 2.

4) Therefore, we will construct a distribution with 11 (the value in the denominator) class intervals.

5) Since we obtained the whole number 2 in step 3, 2 is the number of potential scores to be grouped together in each class interval.

6) We begin the class intervals with a number equal to the lowest number in the distribution, which is zero. Since there are two potential scores in each class interval, our first interval should be 0–1.

7) Skipping to the next number, which is 2, we begin our next class interval. Again, the interval contains two potential scores and takes the form 2–3.

8) Continuing this procedure through eleven class intervals,

we construct the following:

Class Interval
20–21
18–19
16–17
14–15
12–13
10–11
8–9
6–7
4–5
2–3
0–1

9) Do not forget to label the class interval column.

10) The column next to the intervals is labeled "f" for frequency.

11) Counting the number of scores from the ungrouped distribution which fall within each class interval, we obtain the following grouped frequency distribution:

Class Interval	f
20–21	1
18–19	1
16–17	0
14–15	3
12–13	4
10–11	8
8–9	10
6–7	28
4–5	20
2–3	16
0–1	4
	N = 95

Exercise 3. Construct a cumulative frequency distribution from the data in exercise 2.

1) Label a third column "cumulative f."

2) We start with the lowest row of the class interval column. The frequency corresponding to this first class interval ($f = 4$) is placed in the first row of the "cumulative f" column.

3) $16 + 4 = 20$. The sum of the frequency of the second class interval plus the frequency of the first class interval is placed in the second row of the "cumulative f" column.

4) When we continue adding in each successive frequency to the sum as we continue up the column, we obtain the following cumulative frequency distribution:

Class Interval	f	cumulative f
20–21	1	95
18–19	1	94
16–17	0	93
14–15	3	93
12–13	4	90
10–11	8	86
8–9	10	78
6–7	28	68
4–5	20	40
2–3	16	20
0–1	4	4
	$N = 95$	

5) Note that the cumulative frequency of the highest class interval, 95, equals the total number of scores.

2

Pictorial Displays

How to Construct Graphs

Graphs are pictorial displays that organize information into a form that is easy to read. There are several types of graphs. The type of graph that is used to represent data is determined in part by the level at which that variable is measured.

There are four levels of measurement relevant here:

1) *Nominal level measurement.* This is the lowest level of measurement. A nominal-level variable is divided into *categories* that have no mathematical relationship to one another, such as race, political party preference, and so on. The categories cannot be ordered.

2) *Ordinal level measurement.* This is the next highest level of measurement. Here, the categories of a variable can be *ranked, or ordered,* from lowest to highest; for example, leadership ability ranks.

3) *Interval level measurement.* This is the next highest level of measurement. Here, the difference between the same number of units on any part of the scale is equal. We can perform *arithmetic operations* such as addition, subtraction, multiplication and division. Interval scales have *arbitrary zero points* (like temperature readings in which zero degrees still means some temperature).

4) *Ratio level measurement.* This is the highest level of measurement. Here, the difference between each unit of the scale is also uniform, as in interval scales. We can also perform *arithmetic*

operations on ratio level measurement. Unlike interval scales, ratio scales have *true zero points*, (achievement scores, for example, in which a score of zero may mean no measurable achievement at all). The difference between interval scaling and ratio scaling is not important for our purposes. There is no difference in the way they are used in the statistical manipulations reviewed in this manual.

A Technique for Nominal and Ordinal Level Variables: the Bar Graph

Let us suppose we have three classes with 10, 12, and 20 students respectively and we want to represent this visually. To draw a *bar graph*, do the following:

1) Draw a horizontal line and label it with the variable name. Place the categories of the nominal or ordinal variable along the horizontal line. If your variable is nominal, categories can be in any order. If your variable is ordinal, categories should be rank-ordered along the horizontal line.

2) Draw a vertical line perpendicular to the horizontal line. Your vertical line should be three-quarters the length of the horizontal line. Label it "f" (frequency) or "%" (percent of cases), depending upon which is being presented.

3) Place a scale along the vertical line which begins at zero and includes in equally spaced units all of the frequencies or percentages represented in your groups.

4) Place a bar over each category on the horizontal line which reaches as high on the vertical line as the frequency or percentage for that group. These bars should not touch each other.

For our example, the bar graph should look like this:

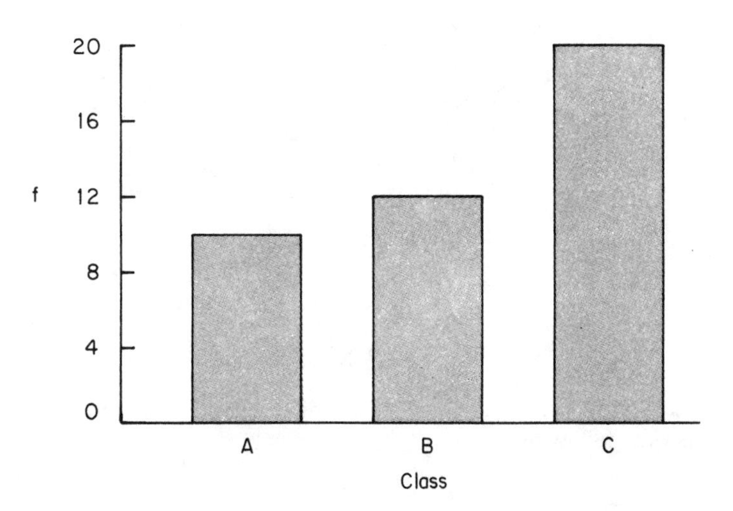

The vertical and horizontal lines can be reversed if you want your bars to be horizontal. You can also place two variables on a bar graph by using sets of adjacent bars of two different colors to represent the two different variables.

Techniques for Interval and Ratio Level Variables

A. The Histogram

A histogram is like a bar graph except that *the bars are usually drawn so that they touch.* Instead of placing categories along the horizontal line, we put our scores or class intervals there. Usually, you will find either class intervals or interval midpoints along the horizontal line.

Suppose we have a frequency distribution like the following:

Score	f
45–49	1
40–44	0
35–39	3
30–34	6
25–29	9
20–24	5
15–19	4
10–14	2
5–9	0
0–4	2
	N = 32

The histogram constructed from this distribution should look like the following:

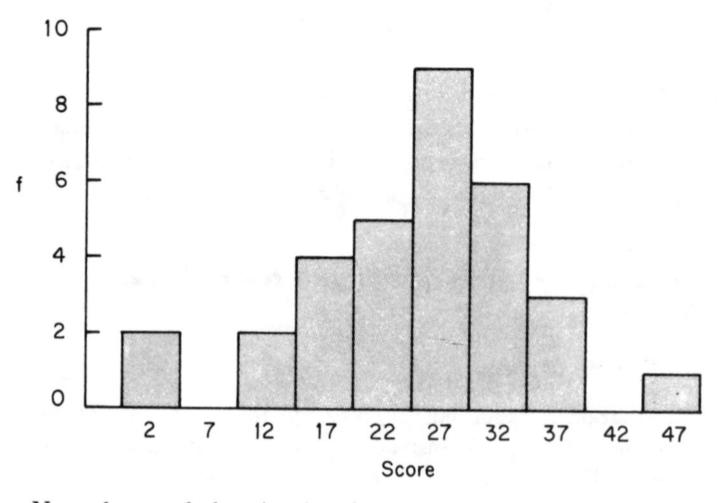

Note that each bar is placed symmetrically over each class interval, so that the interval midpoint is in the *middle* of the bar.

B. The Frequency Polygon.

Rather than a series of bars, the frequency polygon employs *a line of connected dots*, each placed at the proper height for each interval. The horizontal and vertical lines are drawn in the same manner as when constructing a histogram. The frequency polygon constructed from the above frequency distribution should look like the following:

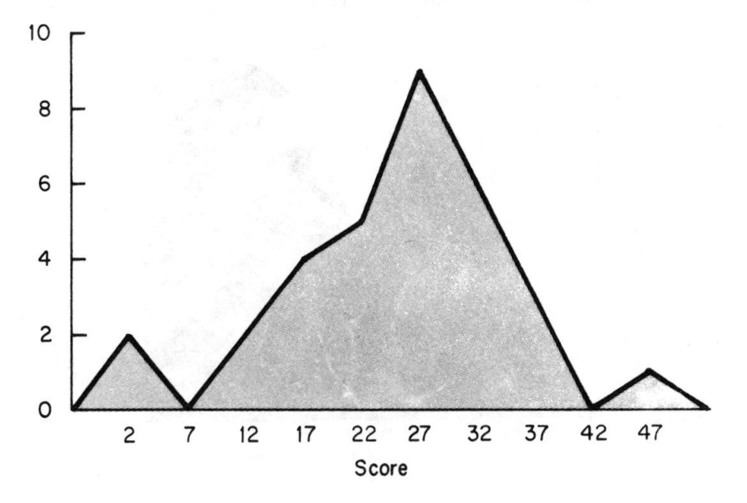

Note that each dot is placed directly over the midpoint for each class interval. Note also that the frequency polygon line is usually drawn back down to meet the horizontal line at both ends.

C. The Cumulative Frequency Curve

Other names for the cumulative frequency curve are: *ogive, s-shaped curve, or percentile curve*. It is constructed in the same way as the frequency polygon except that dots are placed at the upper true limits of each class interval instead of above each midpoint. The upper true limit is equal to the upper value of the class interval plus half a unit. In this case, the upper limit of 0–4 is 4.5; the upper limit of 5–9 is 9.5, and so on. Also, cumulative frequencies are placed along the vertical axis, rather than regular frequencies. A second vertical line of height equal to the first vertical line is often placed on the other end of the horizontal line and cumulative percentages are scaled along the line, usually in units of ten. They are spaced equally throughout the height of the vertical line.

A cumulative frequency curve *always* has the same general "s" shape. One constructed from the frequency distribution in this chapter should look like the following:

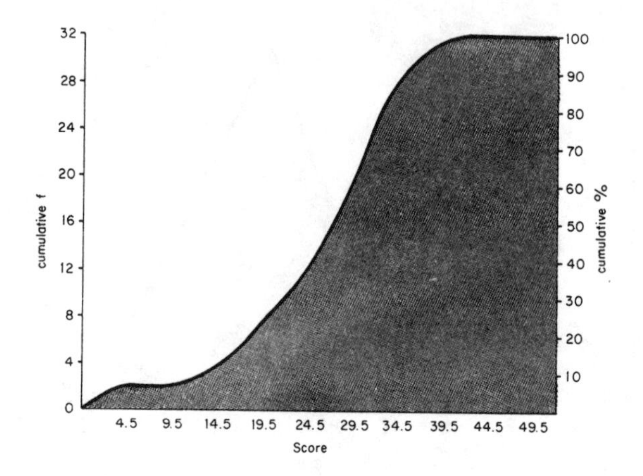

Always Remember

When drawing a graph, always rememebr to:

1) Keep your *vertical line 3/4 the size of your horizontal line.*
2) Label both your horizontal and vertical lines.

Review Questions

1. What are the differences between nominal, ordinal, interval and ratio data?
2. What are two examples of each level of measurement?
3. What are three techniques for pictorially representing interval or ratio data?
4. What is the difference between a histogram and a frequency polygon?
5. What are two other names for a cumulative frequency curve?
6. What are the differences between a bar graph and a histogram?

7. Which line of your graph should be 3/4 the length of the other?

Review Exercises

Exercise 1. Samples were taken of three different types of clerical workers (receptionists, typists and file clerks) to study average number of sick days used by each of the three groups. The receptionists took an average of 8 sick days per year, the typists an average of 9, and the file clerks an average of 11. Represent this information in a bar graph.

1) The horizontal line should be labeled with a variable name, such as "Type of Clerical Worker." the categories receptionist, typist and file clerk should be placed along this horizontal line.

2) The vertical line should be labeled "f" because frequencies are being presented. Remember to make the vertical line three-quarters the length of the horizontal line.

3) The scale along the vertical line should begin at zero and include the frequencies up through 11 in equally spaced units.

4) Bars over each category reach as high as the frequency for that group. Remember that the bars should not touch:

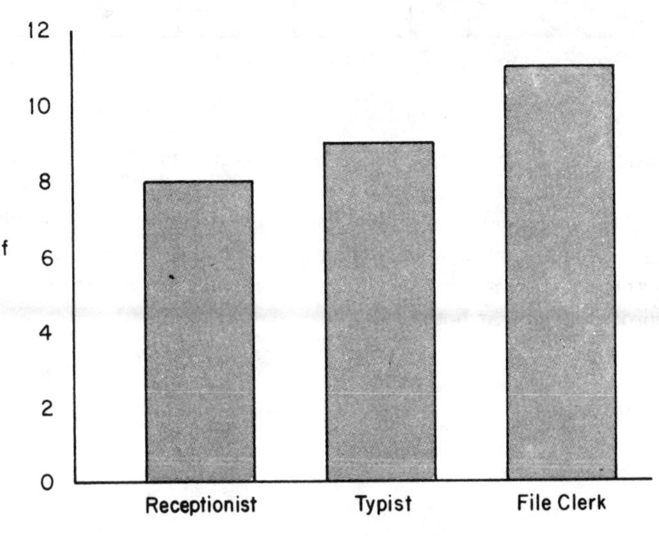

Exercise 2. Construct a histogram for the grouped frequency distribution constructed in Chapter 1, exercise 2.

1) Remember that histograms look like bar graphs, except that the bars usually touch. Note the placement of the bars over the class intervals:

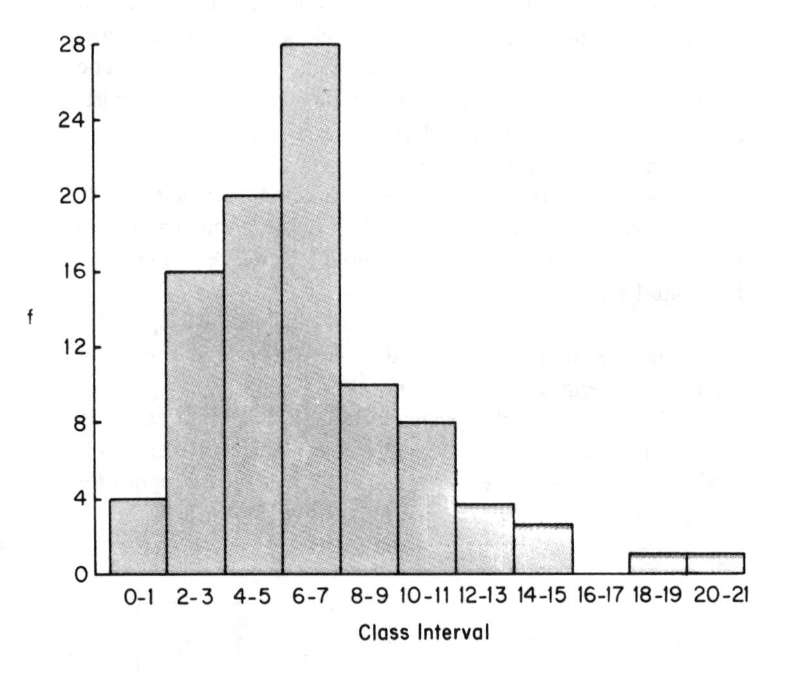

Exercise 3. Construct a frequency polygon for the data in Chapter 1, exercise 1.

1) Remember that frequency polygons use a line of connected dots, rather than bars. Note the placement of the dots and note that the line connecting the dots is drawn back down to meet the horozontal line at both ends:

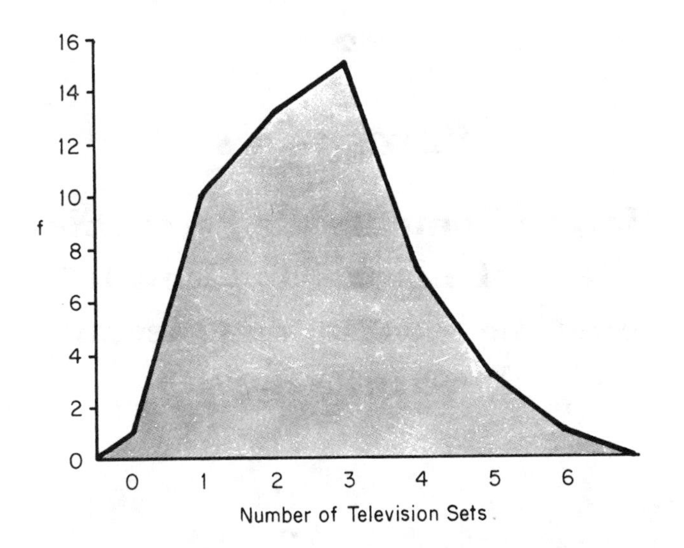

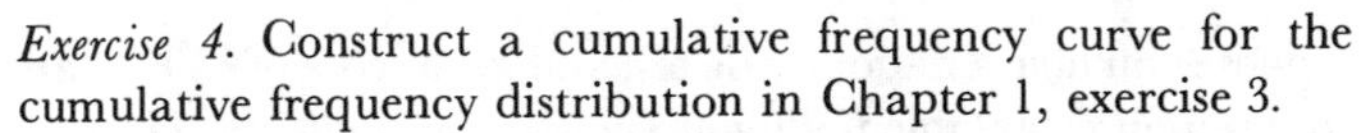

Exercise 4. Construct a cumulative frequency curve for the cumulative frequency distribution in Chapter 1, exercise 3.

1) Remember that a cumulative frequency curve is s-shaped. Note the placement of the dots over the upper true limit of each class interval (upper value plus half a unit):

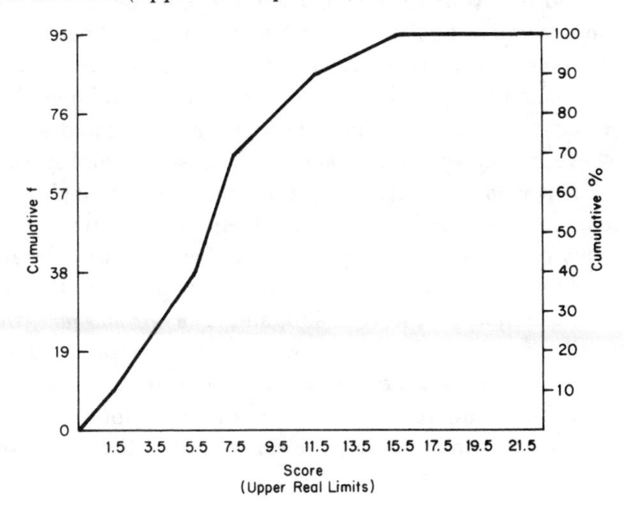

3

Percentiles

How to Compute the Percentile Rank of a Score, and How to Find the Score at a Particular Percentile Rank

Suppose you take a nation-wide test and get a score of 650. That score is uninterpretable until you know the *percentile rank* of your score. What the percentile rank tells you is how you scored compared to everyone else who took the test. For example, you might find that your score of 650 is in the 92nd percentile and that makes you feel good because it tells you that 92 percent of the test-takers scored *at or below your score.* You now know where you're located compared to other people. You also know that since 92 percent of the people who took the test scored at the same level or below you, another 8 percent scored higher than you. The percentage that scored at or below your score plus the percentage that scored above you will equal 100 percent.

In any problem involving percentiles, there are two relevant pieces of information that you can have about a score: (1) *a raw score* (for example, that score of 650) and (2) *a percentile rank* (the percentile corresponding to the raw score, for example the 92nd percentile). One is generally known from the information provided in the problem, and the other must be found. If you have one piece of information, you can find the other. So, if you have a particular raw score, like 650, you can estimate the percentile rank from the frequency distribution of scores. And if you have a specified percentile rank, like the 92nd percentile, you can estimate the raw score corresponding to that percentile rank.

REMEMBER:

A FREQUENCY DISTRIBUTION shows the *number of times each score occurs* and arranges scores, in order, from lowest to highest.

A CUMULATIVE FREQUENCY DISTRIBUTION shows the number of times a score occurs *at or below* a given value.

Situation 1: You know the raw score and want to find the percentile rank.

Let's suppose that you got a score of 650 on a test and that the following frequency distribution represents the scores of everyone else who took the test.

Class Interval	f	cumulative f
751–800	2	98
701–750	3	96
651–700	3	93
601–650	6	90
551–600	10	84
501–550	16	74
451–500	19	58
401–450	17	39
351–400	9	22
301–350	7	13
251–300	4	6
201–250	2	2
	N = 98	

Before you begin, be sure that you make a *cumulative frequency distribution* to accompany your frequency distribution, as you see above.

To find the percentile rank, follow these steps:

1) Find the class interval containing the score whose percentile rank is being estimated. In this example, it is the class interval 601–650. Circle the class interval because you will be referring back to it.

2) Find the lower true limit of the circled class interval. The lower true limit is equal to the lower value of the interval minus half a unit. In this case, $601 - 0.5 = 600.5$.

3) Subtract the lower true limit of the circled class interval from the score whose percentile rank is bεing estimated. In this example, $650 - 600.5 = 49.5$.

4) Multiply the result of step 3 by the number of scores in the class interval you previously circled. You can find this number by looking under the "f" (frequency) column. In this case, $(49.5) \times (6) = 297$.

5) Find the width in true limits of the class interval you previously circled. This is generally one unit more than the distance between the lower and upper numbers in the class interval. In this case, $650 - 601 = 49$, the distance between the lower and upper numbers of the interval. $49 + 1 =$ interval width of 50.

6) Divide the number you calculated in step 4 by the number you computed in step 5. In this case, $297/50 = 5.94$.

7) Add the result to the cumulative frequency corresponding to the class interval which is immediately below the one you circled. In this case, $5.94 + 84 = 89.94$.

8) Divide the result of step 7 by the total number of scores. In this example, $89.94/98 = .9178$.

9) Multiply the result by 100. This will be your percentile rank. In this example, $(.9178) \times (100) = 91.78$. We can round this to the nearest whole percentile of 92.

These steps are based upon the following formula for finding the percentile rank of a raw score:

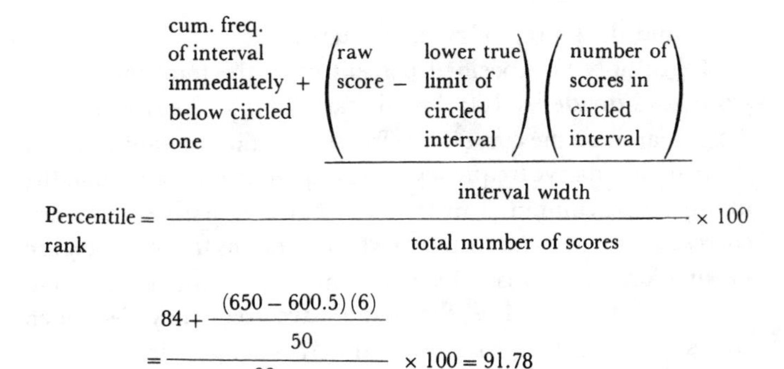

$$\text{Percentile rank} = \frac{\text{cum. freq. of interval immediately below circled one} + \dfrac{\left(\text{raw score} - \text{lower true limit of circled interval}\right)\left(\text{number of scores in circled interval}\right)}{\text{interval width}}}{\text{total number of scores}} \times 100$$

$$= \frac{84 + \dfrac{(650 - 600.5)(6)}{50}}{98} \times 100 = 91.78$$

Situation 2: You have a percentile rank specified and want to find the corresponding raw score, which is unknown.

Let's suppose that you are in charge of admissions to a program and you want to admit only those scoring in the 80th percentile or above on the national exam represented by the frequency distribution in this chapter. You would then have to find the cutoff raw score corresponding to the 80th percentile.

Class Interval	f	cumulative f
751–800	2	98
701–750	3	96
651–700	3	93
601–650	6	90
(551–600	10)	84
501–550	16	74
451–500	19	58
401–450	17	39
351–400	9	22
301–350	7	13
251–300	4	6
201–250	2	2
	N = 98	

Before you begin, be sure to make a *cumulative frequency distribution* to accompany your frequency distribution, as you see above.

To find the score, follow these steps:

1) Multiply the specified percentile by the total number of scores and divide by 100. In this case, $(80) \times (98)/100 = 78.4$.

2) Search up the cumulative frequency column until you find the first cumulative frequency that is equal to or greater than the number you computed in step 1. That cumulative frequency corresponds to the class interval which contains the score you are looking for. In this case, the first cumulative frequency equal to or greater than 78.4 is the cumulative frequency 84 which corresponds to the class interval 551–600. Circle this class interval because you will be referring back to it.

3) Subtract the cumulative frequency corresponding to the class interval immediately below the one you circled from the cumulative frequency obtained in step 1. In this case, $78.4 - 74 = 4.4$.

4) Find the width in true limits of the class interval you previously circled. Remember that this is generally one unit more than the distance between the lower and upper numbers in the class interval. In this case, $600 - 551 = 49$, the distance between the lower and upper numbers of the interval. $49 + 1 =$ interval width of 50.

5) Multiply the result of step 3 by the result of step 4. In this case, $(4.4) \times (50) = 220$.

6) Divide the result of step 5 by the number of scores in the class interval you previously circled. You can find this number by looking under the "f" column ("f" stands for frequencies). In this case, $220/10 = 22.0$.

7) Find the lower true limit of the circled class interval. Remember that the lower true limit is equal to the lower value of the class interval minus half a unit. In this case, $551 - 0.5 = 550.5$.

8) Add the result of step 6 to the result of step 7. In this example, $22.0 + 550.5 = 572.5$. This is your score. We can round this to the nearest whole score of 573.

These steps are based upon the following formula for finding the raw score corresponding to a specified percentile rank:

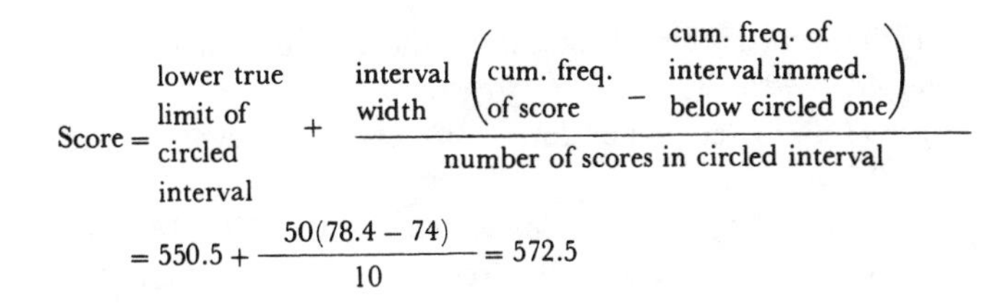

$$\text{Score} = \begin{matrix}\text{lower true}\\ \text{limit of}\\ \text{circled}\\ \text{interval}\end{matrix} + \frac{\text{interval width}\left(\text{cum. freq. of score} - \text{cum. freq. of interval immed. below circled one}\right)}{\text{number of scores in circled interval}}$$

$$= 550.5 + \frac{50(78.4 - 74)}{10} = 572.5$$

Review Questions

1. What does a percentile rank tell you?
2. What information must you have to find the percentile rank of a score? To find the score corresponding to a specified percentile rank?

Review Exercises

Exercise 1. The following table shows the size of 300 classroom groups in a midwestern college:

Size of group	f	cumulative f
30–31	2	300
28–29	0	298
26–27	2	298
24–25	11	296
22–23	13	285
20–21	29	272
18–19	68	243
16–17	64	175
14–15	63	111
12–13	42	48
10–11	5	6
8–9	1	1
	$N = 300$	

Compute the percentile rank of a classroom size of 15.

Remember to make sure that you have a cumulative frequency distribution before you begin.

1) The class interval containing the classroom size of 15 is the interval 14–15.

2) The lower true limit of class interval 14–15 is $14 - 0.5 = 13.5$.

3) The score minus the lower true limit $= 15 - 13.5 = 1.5$.

4) 1.5 multiplied by the frequency of the class interval $14–15 = (1.5) \times (63) = 94.5$.

5) The width of the class interval 14–15 in true limits is 2.

6) The result of step 4 divided by the result of step $5 = 94.5/2 = 47.25$.

7) 47.25 added to the cumulative frequency of the class interval immediately below $14–15 = 47.25 + 48 = 95.25$.

8) 95.25 divided by the total number of scores $= 95.25/300 = .3175$.

9) $(.3175) \times (100) =$ a percentile rank of 31.75.

Exercise 2. Compute the percentile rank of a score of 10 in the frequency distribution based upon Chapter 1, exercises 2 & 3.

1) The class interval containing 10 days of dining out is 10–11.

2) The lower true limit of class interval 10–11 is $10 - 0.5 = 9.5$.

3) The score minus the lower true limit $= 10 - 9.5 = 0.5$.

4) 0.5 multipled by the frequency of the class interval $10–11 = (0.5) \times (8) = 4$.

5) The width of the class interval 10–11 in true limits is 2.

6) The result of step 4 divided by the result of step $5 = 4/2 = 2$.

7) 2 added to the cumulative frequency of the class interval immediately below $10–11 = 2 + 78 = 80$.

8) 80 divided by the total number of scores $= 80/95 = .8421$.

9) $(.8421) \times (100) =$ a percentile rank of 84.21.

Exercise 3. Compute the score occurring at the 75th percentile in the frequency distribution based upon Chapter 1, exercises 2 & 3.

1) 75, the specified percentile, multiplied by the total number of scores and divided by $100 = (75) \times (95)/100 = 71.25$.

2) Searching up the cumulative f column, 78 is the first cumulative frequency that is equal to or greater than 71.25. It corresponds to the class interval 8–9.

3) The cumulative frequency obtained in step 1 minus the cumulative frequency of the class interval immediately below 8–9 = 71.25 – 68 = 3.25.

4) The width of the class interval 8–9 in true limits is 2.

5) The result of step 3 multiplied by the result of step 4 = (3.25) × (2) = 6.50.

6) The result of step 5 divided by the number of scores in the class interval 8–9 = 6.50/10 = .65.

7) The lower true limit of class interval 8–9 is 8 – 0.5 = 7.5.

8) The result of step 6 added to the result of step 7 = .65 + 7.5 = 8.15. This is the score at the 75th percentile.

Exercise 4. Compute the score occurring at the 90th percentile in the frequency distribution in exercise 1 of this chapter.

1) 90, the specified percentile, multiplied by the total number of scores and divided by 100 = (90) × (300)/100 = 270.

2) Searching up the cumulative f column, 272 is the first cumulative frequency that is equal to or greater than 270. It corresponds to the class interval 20–21.

3) The cumulative frequency obtained in step 1 minus the cumulative frequency of the class interval immediately below 20–21 = 270 – 243 = 27.

4) The width of the class interval 20–21 is 2.

5) The result of step 3 multiplied by the result of step 4 = (27) × (2) = 54.

6) The result of step 5 divided by the number of scores in the class interval 20–21 = 54/29 = 1.86.

7) The lower true limit of class interval 20–21 is 20 – 0.5 = 19.5.

8) The result of step 6 added to the result of step 7 = 1.86 + 19.5 = 21.36. This is the score at the 90th percentile.

4

Central Tendency

How to Compute the Mean, the Median, and the Mode

This chapter reviews different ways of describing the "average" score or numerical value.

The Mean

Methods for cumulating the mean vary slightly, depending upon whether your numerical values are in the form of a) raw data, b) an *ungrouped* frequency distribution, or c) a *grouped* frequency distribution.

REMEMBER:

RAW DATA is a *simple list* of scores.

An UNGROUPED FREQUENCY DISTRIBUTION organizes scores or values from lowest to highest and indicates *the number of times each score occurs*.

A GROUPED FREQUENCY DISTRIBUTION combines several scores together into *class intervals* and then organizes the class intervals from lowest to highest. It indicates the number of times a score falls within each class interval.

A. Calculating the Mean from Raw data

Let's suppose the following are a set of scores from a class examination:

81	87	89	92
86	83	88	90
93	89	86	87
90	89	85	88
84	85	87	84
86	87	89	92
90	88	85	87
90	87	89	88
91	86	88	86

To obtain the mean:

1) Add the numbers up. In this case, the sum of the scores is 3152.

2) Divide by the number of scores. In this case, the total number of scores is 36. The mean = 3152/36 = 87.56.

These instructions are based upon the following formula:

$$\text{Mean} = \frac{\text{sum of scores}}{\text{number of scores}} \text{ or, in symbolic form, } \bar{X} = \frac{\sum X}{N} = \frac{3152}{36} = 87.56$$

$\bar{X}$ is the symbol for the *mean* of a sample of scores.
X is the symbol for *score*.
$\sum$ is the symbol for *sum of*.
$\sum X$ is the symbol for *sum of all of the scores*.
N is the symbol for *number of scores*

B. Calculating the Mean from an Ungrouped Frequency Distribution

Let's suppose the above list of scores is organized into an ungrouped frequency distribution like the following:

X	f
93	1
92	2
91	1
90	4
89	5

88	5
87	6
86	5
85	3
84	2
83	1
82	0
81	1
	N = 36

To compute the mean, do the following:

1) Make a third column labeled "fX." Multiply each value in the X (score) column by its corresponding value in the f (frequency) column and place each result in the fX column. In this case, starting with the lowest value, $(81) \times (1) = 81$; $(82) \times (0) = 0$; $(83) \times (1) = 83$; $(84) \times (2) = 168$; $(85) \times (3) = 255$, and so on. Our frequency distribution now looks like this:

X	f	fX
93	1	93
92	2	184
91	1	91
90	4	360
89	5	445
88	5	440
87	6	522
86	5	430
85	3	255
84	2	168
83	1	83
82	0	0
81	1	81
	N = 36	$\sum fX = 3152$

2) Add up the values in the fX column. In this case, they sum to 3152.

3) Divide the result of step 2 by the total number of scores. Remember that in a frequency distribution the total number of scores equals the sum of the f column and *not* the sum of the X column. In this case, the total number of scores (N) = 36. So, 3152/36 = 87.56.

These instructions are based upon the following formula:

$$\text{Mean} = \frac{\text{sum of: (frequency) (score)}}{\text{total number of scores}}$$ or, in symbolic form,

$$\overline{X} = \frac{\sum fX}{N} = \frac{3152}{36} = 87.56$$

C. Calculating the Mean from a Grouped Frequency Distribution

Let's suppose we have a grouped frequency distribution like the following:

Class Interval	f
751–800	2
701–750	3
651–700	3
601–650	6
551–600	10
501–550	16
451–500	19
401–450	17
351–400	9
301–350	7
251–300	4
201–250	2
	N = 98

You have seen this distribution in previous chapters. To compute the mean, do the following:

1) Make a third column labeled "m" or midpoint.* Find the

*We use each midpoint as an average value for each class interval.

midpoint, or middle value, of each class interval and place it in the midpoint column. The easiest way to do this is to add the upper and lower values of the class interval and divide the sum by 2. In this case, the midpoint of the lowest class interval equals $(201 + 250)/2 = 451/2 = 225.5$. The midpoint of the second class interval equals $(251 + 300)/2 = 551/2 = 275.5$, and so on. Your columns should now look like this:

Class Interval	f	m
751–800	2	775.5
701–750	3	725.5
651–700	3	675.5
601–650	6	625.5
551–600	10	575.5
501–550	16	525.5
451–500	19	475.5
401–450	17	425.5
351–400	9	375.5
301–350	7	325.5
251–300	4	275.5
201–250	2	225.5
	N = 98	

2) Make a fourth column labeled "fm" (frequency times midpoint). Multiply each value in the f (frequency) column by its corresponding value in the midpoint column and place each result in the fm column. In this case, starting with the lower class interval, $(225.5) \times (2) = 451$; $(275.5) \times (4) = 1102$; $(325.5) \times (7) = 2278.5$, etc. Your columns should now look like this:

Class Interval	f	m	fm
751–800	2	775.5	1551.0
701–750	3	725.5	2176.5
651–700	3	675.5	2026.5

601–650	6	625.5	3753.0
551–600	10	575.5	5755.0
501–550	16	525.5	8408.0
451–500	19	475.5	9034.5
401–450	17	425.5	7233.5
351–400	9	375.5	3379.5
301–350	7	325.5	2278.5
251–300	4	275.5	1102.0
201–250	2	225.5	451.0
	N = 98		$\sum$fm = 47,149.0

3) Add up the values in the fm column. In this case, they sum to 47,149.0.

4) Divide the result of step 3 by the total number of scores. Remember that in a frequency distribution the total number of scores equals the sum of the f column and *not* the number of intervals. In this case, the total number of scores (N) = 98. 47,149/98 = 481.11.

These instructions are based upon the following formula:

$$\text{Mean} = \frac{\text{sum of: (frequency)(midpoint of interval)}}{\text{total number of scores}}$$

or, in symbolic form, $\bar{X} = \frac{\sum \text{fm}}{\text{N}} = \frac{47{,}149}{98} = 481.11$

D. Calculating a Weighted Mean

Weighted means are for calculating the overall mean of several different means which were obtained from groups *with unequal numbers of scores in them*. Each individual mean must be weighted by the number in each group to obtain an overall mean.

Suppose we have three groups with the following mean exam scores: 86, 79, 91. Suppose that there are 10 students in the first class, 12 in the second class, and 20 in the class whose mean was 91.

To obtain a weighted mean, do the following:

1) Make a set of four columns which look like this:

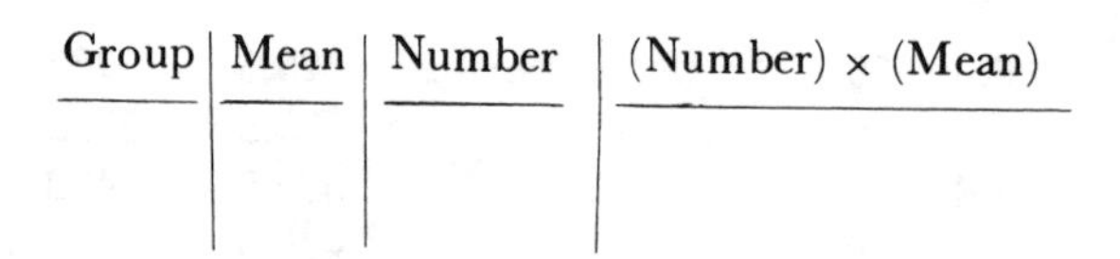

2) List the mean and number of people in each group in the columns like this:

Group	Mean	Number	(Number) × (Mean)
1	86	10	
2	79	12	
3	91	20	

3) Multiply each number in the mean column by the corresponding value in the number column. In this case, (86) × (10) = 860, etc. Your columns should now look like this:

Group	Mean	Number	(Number) × (Mean)
1	86	10	860
2	79	12	948
3	91	20	1820
		N = 42	$\sum = 3628$

4) Add the numbers in the (Number) × (Mean) column. In this case, 860 + 948 + 1820 = 3628.

5) Add the numbers in the number column. In this case, 10 + 12 + 20 = 42.

6) Divide the result of step 4 by the result of step 5. In this case, 3628/42 = 86.38. This is your weighted mean.

The Median

Methods for computing the median vary, depending upon whether your numerical values are in the form of raw data or a frequency distribution.

A. Calculating the Median from Raw Data

Let us look at the raw data on p. 1. To find the median we must first *order all scores* from lowest to highest:

93
92
92
91
90
90
90
90
89
89
89
89
89
88
88
88
88
88
87
87
87
87
87
87
86
86
86
86
86
85
85
85
84

84
83
81

The median of an ordered list of scores is the *centermost score*. If there is an odd number of scores in the list, there will be an exact center. If there is an even number of scores in the list, as in this case, the median is the average of the two centermost scores. In this case, the two centermost scores are 87 and 88. There are 17 scores on either side of these two center scores. The average of these two center scores and therefore the median is 87.5.

B. Calculating the Median from a Frequency Distribution

The method for computing the median from a grouped or ungrouped frequency distribution is similar to the method for finding a raw score when you have a specified percentile rank (see Chapter 3) since *the median is the same as the 50th percentile*. Let us use the grouped frequency distribution on p. 25 as an illustration. Before you begin, make sure that you make a *cumulative frequency distribution* to accompany your frequency distribution:

REMEMBER:
A CUMULATIVE FREQUENCY DISTRIBUTION shows the number of times a score occurs *at or below* each score or class interval, and arranges scores or class intervals in order from lowest to highest.

Class Interval	f	cumulative f
751–800	2	98
701–750	3	96
651–700	3	93
601–650	6	90
551–600	10	84
501–550	16	74
451–500	19	58

401–450	17	39
351–400	9	22
301–350	7	13
251–300	4	6
201–250	2	2
	N = 98	

To find the median, follow these steps:

1) Divide the total number of scores by 2. In this case, 98/2 = 49.

2) Search up the cumulative frequency column until you find the first cumulative frequency that is equal to or greater than the number you computed in step 1. Circle the class interval, or score (X) in the case of an ungrouped frequency distribution, that cumulative frequency corresponds to. In this case, the first cumulative frequency equal to or greater than 49 is 58 which corresponds to a class interval of 451–500.

3) Subtract the cumulative frequency corresponding to the class interval (or score) immediately below the one you circled from the cumulative frequency obtained in step 1. In this case, 49 – 39 = 10.

4) Find the width in true limits of the class interval you previously circled. Remember that this is generally one unit more than the distance between the lower and upper numbers in the class interval. If you have scores, rather than class intervals, the width in true limits of each score is always 1. In this case, 500 – 451 = 49, the distance between the lower and upper numbers of the interval. 49 + 1 = interval width of 50.

5) Multiply the result of step 3 by the result of step 4. In this case, (10) × (50) = 500.

6) Divide the result of step 5 by the number of scores in the class interval you previously circled. You can find this number by looking under the f (frequency) column. In this case, 500/19 = 26.32.

7) Find the lower true limit of the circled class interval. Remember that the lower true limit is equal to the lower value of the class interval minus half a unit. In this case, 451 – 0.5 = 450.5.

8) Add the result of step 6 to the result of step 7. In this example, $26.32 + 450.5 = 476.82$.

These steps are based upon the following formula for finding the median of a frequency distribution:

$$\text{Median} = \begin{matrix}\text{lower true}\\ \text{limit of}\\ \text{circled}\\ \text{score or}\\ \text{interval}\end{matrix} + \frac{\begin{matrix}\text{interval}\\ \text{or score}\\ \text{width}\end{matrix}\left(\dfrac{\text{total number of scores}}{2} - \begin{matrix}\text{cum. freq. of}\\ \text{score or interval}\\ \text{immed. below}\\ \text{circled one}\end{matrix}\right)}{\text{frequency in circled score or interval}}$$

In the case of our example,

$$\text{median} = 450.5 + \frac{50\left(\frac{98}{2} \quad 39\right)}{19} = 476.82$$

Notice that the median we computed is slightly less than the mean we computed for this grouped frequency distribution. When our mean is higher than our median, it indicates that there are some extreme high scores pulling up the mean. This is called a *positively skewed distribution*. When our mean is lower than our median, it indicates that there are some extreme low scores pulling down the mean. This is called a *negatively skewed distribution*. Generally, if there is a lot of skewness, the median is a better indicator of the "average" than is the mean. If our mean and median are equal, our distribution is *symmetrical* and looks something like this:

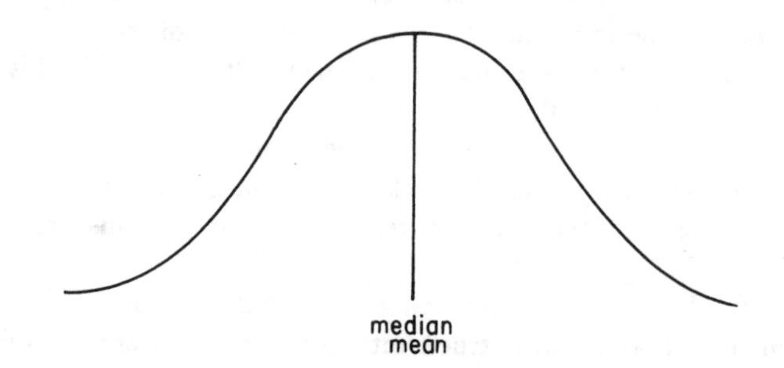

This is also called a *normal distribution* or a *bell-shaped curve*.

The Mode

The mode is the *most frequent score*.

In a simple list of scores, it is the score that is repeated the most (87 in the list in this chapter).

In an ungrouped frequency distribution, the mode is the score with the largest number in the f column (87 in the ungrouped frequency distribution in this chapter).

In a grouped frequency distribution, the mode is the midpoint of the class interval with the largest number in the f column (475.5 in the grouped frequency distribution in this chapter).

The mode is not used as commonly as the mean and median.

Review Questions

1. What is the difference between the mean, median, and mode?
2. When is the median a better indication of the "average" than the mean?
3. What is a skewed distribution? Positively skewed? Negatively skewed?
4. What does a normal distribution look like? What is another term for normal distribution?
5. What is a weighted mean?
6. What percentile rank does the median have?

Review Exercises

Exercise 1. The following are a set of scores on an aptitude exam given at the personnel office of a large corporation:

100
90
80
75
50
70
85

85
65
95

Calculate the mean.

1) The numbers sum to 795.

2) 795 divided by the number of scores = 795/10 = a mean of 79.5.

Exercise 2. The following are the ages at retirement in a random sample of thirty executives:

X	f
64	2
63	7
62	5
61	4
60	5
59	3
58	2
57	0
56	1
55	1
	N = 30

Calculate the mean.

1) We make a third column labeled "fX" and multiply each age value by its corresponding value in the f column to obtain the following:

X	f	fX
64	2	128
63	7	441
62	5	310
61	4	244

60	5	300
59	3	177
58	2	116
57	0	0
56	1	56
55	1	55
	N = 30	$\sum fX = 1827$

2) The sum of the values in the fX column = 1827.

3) 1827 divided by the total number of scores = 1827/30 = a mean of 60.9.

Exercise 3. Calculate the mean of the frequency distribution in Chapter 1, exercise 2.

1) We make a third column labeled "m" for midpoint and list the middle value of each interval:

Class Interval	f	m
20–21	1	20.5
18–19	1	18.5
16–17	0	16.5
14–15	3	14.5
12–13	4	12.5
10–11	8	10.5
8–9	10	8.5
6–7	28	6.5
4–5	20	4.5
2–3	16	2.5
0–1	4	0.5
	N = 95	

2) We make a fourth column labeled "fm" and multiply each value in the f column by its corresponding value in the m column:

Class Interval	f	m	fm
20–21	1	20.5	20.5
18–19	1	18.5	18.5
16–17	0	16.5	0
14–15	3	14.5	43.5
12–13	4	12.5	50.0
10–11	8	10.5	84.0
8–9	10	8.5	85.0
6–7	28	6.5	182.0
4–5	20	4.5	90.0
2–3	16	2.5	40.0
0–1	4	0.5	2.0
	N = 95		$\sum = 615.5$

3) The sum of the values in the fm column = 615.5.

4) 615.5 divided by the total number of scores = 615.5/95 = a mean of 6.48.

Exercise 4. The mean age of 20 vice presidents in corporation A is 45.7; the mean age of 15 vice presidents in corporation B is 43.2, and the mean age of 25 vice presidents in corporation C is 36.8. What is the mean age of all of these vice presidents?

1) We make a set of column headings like the following.

Group	Mean	Number	(Number) × (Mean)

2) The name of the group, the mean, and the number of people in each group are listed in the first three columns:

Group	Mean	Number	(Number) × (Mean)
A	45.7	20	
B	43.2	15	
C	36.8	25	

3) We multiply each number in the mean column by the corresponding value in the number column:

Group	Mean	Number	(Number) × (Mean)
A	45.7	20	914
B	43.2	15	648
C	36.8	25	920
		N = 60	$\sum = 2482$

4) The sum of the values in the (Number) × (Mean) column = 2482.

5) The sum of numbers in the number column = 60.

6) The result of step 4 divided by the result of step 5 = 2482/60 = a weighted mean of 41.37.

Exercise 5. Calculate the median of the following scores:

94
93
88
87
86
85
83
71
70
65

Remember to order all scores first from lowest to highest if they are not already in order. There is an even number of scores in this list. The median is the average of the two centermost scores of 85 and 86. Therefore, the median is 85.5.

Exercise 6. Calculate the median of the frequency distribution in Chapter 1, exercises 2 & 3.

1) The total number of scores divided by 2 = 95/2 = 47.5.

2) Looking up the cumulative frequency column, 68 is the

first cumulative frequency that is equal to or greater than 47.5. It corresponds to the class interval 6–7.

3) The cumulative frequency obtained in step 1 minus the cumulative frequency of the class interval immediately below 6–7 = 47.5 – 40 = 7.5.

4) The width in true limits of the class interval 6–7 is 2.

5) The result of step 3 multiplied by the result of step 4 = (7.5) × (2) = 15.

6) The result of step 5 divided by the frequency of the class interval 6–7 = 15/28 = .54.

7) The lower true limit of the class interval 6–7 is 6 – 0.5 = 5.5.

8) The result of step 6 added to the result of step 7 = .54 + 5.5 = a median of 6.04.

Exercise 7. What is the mode of the frequency distribution in Chapter 1, exercise 2?

The mode is the most frequent score. In a grouped frequency distribution, it is the midpoint of the class interval with the largest number in the f column. In this frequency distribution, the mode is 6.5.

5

Measures of Dispersion

How to Compute the Standard Deviation and the Variance

The *standard deviation* is the most common way of describing the way scores disperse themselves around the mean. If there is a lot of spread of scores around the mean, the standard deviation will be comparatively large, relative to the mean. If there is only a small degree of spread of scores around the mean, the standard deviation will be comparatively small, relative to the mean. *The standard deviation is always interpreted with reference to the mean.*

One standard deviation indicates how many units you must go either below or above the mean to include 34.13% of all scores (assuming the distribution is *normal* and therefore *symmetrical*).

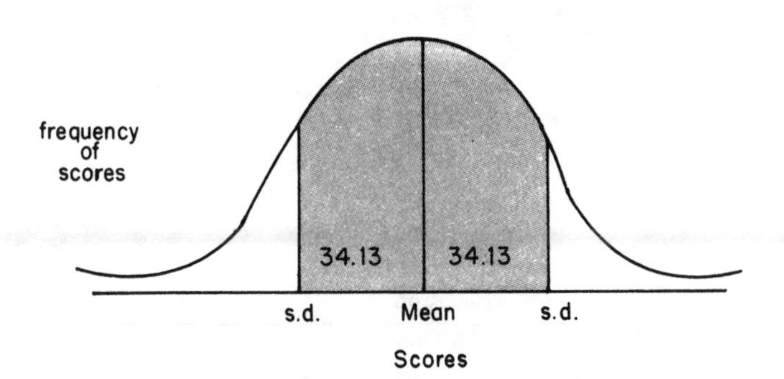

Between one standard deviation below the mean and one standard deviation above the mean occur 68.26% of all scores (34.13 + 34.13 = 68.26). If, for example, the mean is 90 in a normal distribution and the standard deviation is 10, then 68.26% of all scores should fall between 80 and 100 (mean – 1 s.d. = 90 – 10 units = 80; mean + 1 s.d. = 90 + 10 units = 100).

The *variance* is the *standard deviation squared.* If the standard deviation of a distribution of scores is 10, then the variance is 100 ($(10)^2 = (10) \times (10) = 100$). You can see some uses of the standard deviation and variance in Chapters 6, 8, 9, 10 and 13.

Methods for computing the standard deviation vary slightly, depending upon whether your numerical values are in the form of raw data, an ungrouped frequency distribution, or a grouped frequency distribution. In order to compute a standard deviation, you should already know how to compute a mean. Please refer to Chapter 4.

A. Calculating the Standard Deviation from Raw Data

Let's look at the raw data from which you computed the mean in Chapter 4. These are a set of scores from a class examination:

81
86
93
90
84
86
90
90
91
87
83
89
89
85
87
88

87
86
89
88
86
85
87
89
85
89
88
92
90
87
88
84
92
87
88
86

To obtain the standard deviation, do the following:

1) Line up all of your scores in a column such as you see above and label this column "X" or "score".

2) Make a second column and label it "$(X - \overline{X})$" or "deviations from the mean." Subtract the mean from each score in the X column. The mean has already been calculated for this list of scores in Chapter 4. It equals 87.56. In this case, $81 - 87.56 = -6.56$; $86 - 87.56 = -1.56$; $93 - 87.56 = 5.44$; etc.

Your scores should now look like this:

X	$(X - \overline{X})$
81	– 6.56
86	– 1.56
93	5.44
90	2.44

84	– 3.56
86	– 1.56
90	2.44
90	2.44
91	3.44
87	– 0.56
83	– 4.56
89	1.44
89	1.44
85	– 2.56
87	– 0.56
88	0.44
87	– 0.56
86	– 1.56
89	1.44
88	0.44
86	– 1.56
85	– 2.56
87	– 0.56
89	1.44
85	– 2.56
89	1.44
88	0.44
92	4.44
90	2.44
87	– 0.56
88	0.44
84	– 3.56
92	4.44
87	– 0.56
88	0.44
86	– 1.56

3) Make a third column and label it "$(X - \overline{X})^2$" or "deviations squared". Square each value in the $(X - \overline{X})$ column. For example, – 6.56 squared = 43.03.

Your columns should now look like this:

X	$(X - \overline{X})$	$(X - \overline{X})^2$
81	– 6.56	43.03
86	– 1.56	2.43
93	5.44	29.59
90	2.44	5.95
84	– 3.56	12.67
86	– 1.56	2.43
90	2.44	5.95
90	2.44	5.95
91	3.44	11.83
87	– 0.56	0.31
83	– 4.56	20.79
89	1.44	2.07
89	1.44	2.07
85	– 2.56	6.55
87	– 0.56	0.31
88	0.44	0.19
87	– 0.56	0.31
86	– 1.56	2.43
89	1.44	2.07
88	0.44	0.19
86	– 1.56	2.43
85	– 2.56	6.55
87	– 0.56	0.31
89	1.44	2.07
85	– 2.56	6.55
89	1.44	2.07
88	0.44	0.19
92	4.44	19.71
90	2.44	5.95
87	– 0.56	0.31
88	0.44	0.19
84	– 3.56	12.67

92	4.44	19.71
87	– 0.56	0.31
88	0.44	0.19
86	– 1.56	2.43
		$\sum = 238.76$

4) Add up the values in the $(X - \overline{X})^2$ column. In this case they sum to 238.76.

5) Divide the result of step 4 by the total number of scores. In this case, 238.76/36 = 6.63. This is the variance for this list of scores.

6) Compute the square root of the result of step 5. The easiest way to do this is with a hand calculator that has a square root function. It looks like this: $\sqrt{\ }$. The square root of a number is the value which must be multiplied by itself to obtain the original number. In this case, the square root of 6.63 is 2.57. This is the standard deviation.

The above instructions are based upon the following formula:

$$\text{standard deviation} = \sqrt{\frac{\text{sum of: squared (score – mean)}}{\text{total number of scores}}}$$

or, in symbolic form,

$$s = \sqrt{\frac{\Sigma(X - \overline{X})^2}{N}} = \sqrt{\frac{238.76}{36}} = 2.57$$

B. Calculating the Standard Deviation from an Ungrouped Frequency Distribution

Let's suppose our scores are organized into an ungrouped frequency distribution:

X	f
93	1
92	2
91	1
90	4
89	5

88	5
87	6
86	5
85	3
84	2
83	1
82	0
81	1
	N = 36

To obtain the standard deviation do the following:

1) Make a third column and label it $(X - \overline{X})$ or "deviations from the mean." Subtract the mean from each score in the X column. The mean has already been calculated for this frequency distribution in Chapter 4. It equals 87.56. In this case, $93 - 87.56 = 5.44$; $92 - 87.56 = 4.44$; etc.

Your scores should now look like this:

X	f	$(X - \overline{X})$
93	1	5.44
92	2	4.44
91	1	3.44
90	4	2.44
89	5	1.44
88	5	0.44
87	6	− 0.56
86	5	− 1.56
85	3	− 2.56
84	2	− 3.56
83	1	− 4.56
82	0	− 5.56
81	1	− 6.56
	N = 36	

2) Make a fourth column and label it $(X - \overline{X})^2$ or "squared deviations." Square each value in the $(X - \overline{X})$ column. For

example, – 6.56 squared = 43.03. Your columns should now look like this:

X	f	$(X - \overline{X})$	$(X - \overline{X})^2$
93	1	5.44	29.59
92	2	4.44	19.71
91	1	3.44	11.83
90	4	2.44	5.95
89	5	1.44	2.07
88	5	0.44	0.19
87	6	– 0.56	0.31
86	5	– 1.56	2.43
85	3	– 2.56	6.55
84	2	– 3.56	12.67
83	1	– 4.56	20.79
82	0	– 5.56	30.91
81	1	– 6.56	43.03
	$N = 36$		

3) Make a fifth column and label it "$f(X - X)^2$" or "frequency times squared deviations." Multiply each value in the f column by its corresponding value in the $(X - \overline{X})^2$ column; e.g., $(29.59) \times (1) = 29.59$; $(19.71) \times (2) = 39.42$; etc.

Your columns should now look like this:

X	f	$(X - X)$	$(X - \overline{X})^2$	$f(X - \overline{X})^2$
93	1	5.44	29.59	29.59
92	2	4.44	19.71	39.42
91	1	3.44	11.83	11.83
90	4	2.44	5.95	23.80
89	5	1.44	2.07	10.35
88	5	0.44	0.19	0.95
87	6	– 0.56	0.31	1.86
86	5	– 1.56	2.43	12.15
85	3	– 2.56	6.55	19.65
84	2	– 3.56	12.67	25.34

83	1	– 4.56	20.79	20.79
82	0	– 5.56	30.91	0
81	1	– 6.56	43.03	43.03
	N = 36			$\sum = 238.76$

4) Add up the values in the $f(X - \overline{X})^2$ column. In this case they sum to 238.76.

5) Divide the result of step 4 by the total number of scores. In this case, 238.76/36 = 6.63. This is the variance for this frequency distribution.

6) Compute the square root of the result of step 5. The square root of 6.63 is 2.57. This is the standard deviation.

The above instructions are based upon the following formula:

$$\text{standard deviation} = \sqrt{\frac{\text{sum of: (frequency) [squared (score – mean)]}}{\text{total number of scores}}}$$

or, in symbolic form,

$$s = \sqrt{\frac{\sum f(X - \overline{X})^2}{N}} = \sqrt{\frac{238.76}{36}} = 2.57$$

C. Calculating the Standard Deviation from a Grouped Frequency Distribution

Let's continue to work with the grouped frequency distribution you last saw in Chapter 4. To compute the mean, we have already added a third midpoint column:

Class Interval	f	m
751–800	2	775.5
701–750	3	725.5
651–700	3	675.5
601–650	6	625.5
551–600	10	575.5
501–550	16	525.5
451–500	19	475.5
401–450	17	425.5
351–400	9	375.5

301–350	7	325.5
251–300	4	275.5
201–250	2	225.5
	N = 98	

1) Make a fourth column and label it "$(m - \overline{X})$" or "deviations from the mean."* Subtract the mean from each midpoint in the m column. The mean has already been calculated for this grouped frequency distribution in Chapter 4. It equals 481.11. In this case, 775.5 – 481.11 = 294.39; 725.5 – 481.11 = 244.39; etc.

Your columns should now look like this:

Class Interval	f	m	$(m - \overline{X})$
751–800	2	775.5	294.39
701–750	3	725.5	244.39
651–700	3	675.5	194.39
601–650	6	625.5	144.39
551–600	10	575.5	94.39
501–550	16	525.5	44.39
451–500	19	475.5	– 5.61
401–450	17	425.5	– 55.61
351–400	9	375.5	– 105.61
301–350	7	325.5	– 155.61
251–300	4	275.5	– 205.61
201–250	2	225.5	– 255.61
	N = 98		

2) Make a fifth column and label it "$(m - \overline{X})^2$" or "squared deviations." For example, 294.34 squared = 86,665.47; 244.39 squared = 59,726.47; etc. Your columns should now look like this:

*We use each midpoint as an average value for each class interval.

Class Interval	f	m	$(m - \bar{X})$	$(m - \bar{X})^2$
751–800	2	775.5	294.39	86,665.47
701–750	3	725.5	244.39	59,726.47
651–700	3	675.5	194.39	37,787.47
601–650	6	625.5	144.39	20,848.47
551–600	10	575.5	94.39	8,909.47
501–550	16	525.5	44.39	1,970.47
451–500	19	475.5	– 5.61	31.47
401–450	17	425.5	– 55.61	3,092.47
351–400	9	375.5	– 105.61	11,153.47
301–350	7	325.5	– 155.61	24,214.47
251–300	4	275.5	– 205.61	42,275.47
201–250	2	225.5	– 255.61	65,336.47
	N = 98			

3) Make a sixth column and label it $f(m - \bar{X})^2$ or "frequency times squared deviations". Multiply each value in the f column by its corresponding value in the $(m - \bar{X})^2$ column.

To see how your columns should now look, turn to p. 58.

4) Add up the values in the $f(m - \bar{X})^2$ column. They sum to 1,334,412.90.

5) Divide the result of step 4 by the total number of scores. In this case, 1,334,412.90/98 = 13,616.46. This is the variance for this frequency distribution.

6) Compute the square root of the result of step 5. The square root of 13,616.46 is 116.69. This is the standard deviation.

The above instructions are based upon the following formula:

$$\text{standard deviation} = \sqrt{\frac{\text{sum of: (frequency) [squared (midpoint – mean)]}}{\text{total number of scores}}}$$

or, in symbolic form,

$$s = \sqrt{\frac{\Sigma f(m - \bar{X})^2}{N}} = \sqrt{\frac{1{,}334{,}412.90}{98}} = 116.69$$

Class Interval	f	m	$(m - \overline{X})$	$(m - \overline{X})^2$	$f(m - \overline{X})^2$
751–800	2	775.5	294.39	86,665.47	173,330.94
701–750	3	725.5	244.39	59,726.47	179,179.41
651–700	3	675.5	194.39	37,787.47	113,362.41
601–650	6	625.5	144.39	20,848.47	125,090.82
551–600	10	575.5	94.39	8,909.47	89,094.70
501–550	16	525.5	44.39	1,970.47	31,527.52
451–500	19	475.5	– 5.61	31.47	597.93
401–450	17	425.5	– 55.61	3,092.47	52,571.99
351–400	9	375.5	– 105.61	11,153.47	100,381.23
301–350	7	325.5	– 155.61	24,214.47	169,501.29
251–300	4	275.5	– 205.61	42,275.47	169,101.88
201–250	2	225.5	– 255.61	65,336.47	130,672.94
	$N = 98$				$\sum = 1,334,412.90$

Review Questions

1. What does the standard deviation describe?
2. The standard deviation is interpreted with reference to what other measure?
3. If you know the standard deviation, how can you find the variance?

Review Exercises

Exercise 1. Calculate the standard deviation of the scores in Chapter 4, exercise 1.

1) If not already lined up in a column, the scores should be placed in a column labeled "X."

2) We make a second column labeled $(X - \overline{X})$ and subtract the mean from each value in the X column. The mean has already been calculated for this list of scores in Chapter 4. It equals 79.5. We obtain the following:

X	$(X - \overline{X})$
100	20.5
90	10.5
80	0.5
75	− 4.5
50	− 29.5
70	− 9.5
85	5.5
85	5.5
65	− 14.5
95	15.5

3) We square each value in the $(X - \overline{X})$ column:

X	$(X - \overline{X})$	$(X - \overline{X})^2$
100	20.5	420.25
90	10.5	110.25
80	0.5	0.25
75	− 4.5	20.25

50	– 29.5	870.25
70	– 9.5	90.25
85	5.5	30.25
85	5.5	30.25
65	– 14.5	210.25
95	15.5	240.25
		$\sum = 2022.50$

4) The sum of the values in the $(X - \overline{X})^2$ column = 2022.50.

5) The result of step 4 divided by the total number of scores = 2022.50/10 = 202.25.

6) The square root of the result of step 5 = $\sqrt{202.25} = 14.22$. This is the standard deviation.

Exercise 2. Calculate the standard deviation of the scores in Chapter 4, exercise 2.

We use the X and f columns from when we calculated a mean of 60.9 for this distribution in Chapter 4.

1) We make a third column labeled $(X - \overline{X})$ and subtract the mean from each score in the X column:

X	f	$(X - \overline{X})$
64	2	3.1
63	7	2.1
62	5	1.1
61	4	0.1
60	5	– 0.9
59	3	– 1.9
58	2	– 2.9
57	0	– 3.9
56	1	– 4.9
55	1	– 5.9
	N = 30	

2) We make a fourth column labeled $(X - \overline{X})^2$ and square each value in the $(X - \overline{X})$ column:

X	f	$(X - \bar{X})$	$(X - \bar{X})^2$
64	2	3.1	9.61
63	7	2.1	4.41
62	5	1.1	1.21
61	4	0.1	0.01
60	5	– 0.9	0.81
59	3	– 1.9	3.61
58	2	– 2.9	8.41
57	0	– 3.9	15.21
56	1	– 4.9	24.01
55	1	– 5.9	34.81
	N = 30		

3) We multiply each value in the f column by its corresponding value in the $(X - \bar{X})^2$ column:

X	f	$(X - \bar{X})$	$(X - \bar{X})^2$	$f(X - \bar{X})^2$
64	2	3.1	9.61	19.22
63	7	2.1	4.41	30.87
62	5	1.1	1.21	6.05
61	4	0.1	0.01	0.04
60	5	– 0.9	0.81	4.05
59	3	– 1.9	3.61	10.83
58	2	– 2.9	8.41	16.82
57	0	– 3.9	15.21	0
56	1	– 4.9	24.01	24.01
55	1	– 5.9	34.81	34.81
	N = 30			$\sum = 146.70$

4) The sum of the values in the $f(X - \bar{X})^2$ column = 146.70.

5) The result of step 4 divided by the total number of scores = 146.7/30 = 4.89.

6) The square root of the result of step 5 = $\sqrt{4.89} = 2.21$. This is the standard deviation.

Exercise 3. Calculate the standard deviation of the grouped frequency distribution constructed in Chapter 1, exercise 2.

We use the first three columns from when we calculated a mean of 6.48 for this distribution in Chapter 4: the Class Interval, f and m columns.

1) We make a fourth column labeled $(m - \overline{X})$ and subtract the mean from each midpoint:

Class Interval	f	m	$(m - \overline{X})$
20–21	1	20.5	14.02
18–19	1	18.5	12.02
16–17	0	16.5	10.02
14–15	3	14.5	8.02
12–13	4	12.5	6.02
10–11	8	10.5	4.02
8–9	10	8.5	2.02
6–7	28	6.5	0.02
4–5	20	4.5	– 1.98
2–3	16	2.5	– 3.98
0–1	4	0.5	– 5.98
	N = 95		

2) We make a fifth column labeled $(m - \overline{X})^2$ and square each value in the $(m - \overline{X})$ column:

Class Interval	f	m	$(m - \overline{X})$	$(m - \overline{X})^2$
20–21	1	20.5	14.02	196.56
18–19	1	18.5	12.02	144.48
16–17	0	16.5	10.02	100.40
14–15	3	14.5	8.02	64.32
12–13	4	12.5	6.02	36.24
10–11	8	10.5	4.02	16.16
8–9	10	8.5	2.02	4.08
6–7	28	6.5	0.02	0.00
4–5	20	4.5	– 1.98	3.92
2–3	16	2.5	– 3.98	15.84
0–1	4	0.5	– 5.98	35.76
	N = 95			

3) We make a sixth column labeled $f(m - \overline{X})^2$ and multiply each value in the f column by its corresponding value in the $(m - \overline{X})^2$ column:

Class Interval	f	m	$m - \overline{X}$	$(m - \overline{X})^2$	$f(m - \overline{X})^2$
20–21	1	20.5	14.02	196.56	196.56
18–19	1	18.5	12.02	144.48	144.48
16–17	0	16.5	10.02	100.40	0
14–15	3	14.5	8.02	64.32	192.96
12–13	4	12.5	6.02	36.24	144.96
10–11	8	10.5	4.02	16.16	129.28
8–9	10	8.5	2.02	4.08	40.80
6–7	28	6.5	0.02	0.00	0
4–5	20	4.5	− 1.98	3.92	78.40
2–3	16	2.5	− 3.98	15.84	253.44
0–1	4	0.5	− 5.98	35.76	143.04
	N = 95				Σ = 1323.92

4) The sum of the values in the $f(m - \overline{X})^2$ column = 1323.92.

5) The result of step 4 divided by the total number of scores = 1323.92/95 = 13.94.

6) The square root of the result of step 5 = $\sqrt{13.94} = 3.73$. This is the standard deviation.

6

The Normal Distribution

How to Standardize Scores, and How to Locate the Proportion of Cases Falling Between Two Scores

Remember that in Chapter 5 you learned that a *standard deviation* indicates how many units you must go either below or above the mean to include 34.13% of all scores in a *normal distribution.* The percentage of scores between the mean and one standard deviation *is always the same* in a normal distribution. The percentage of scores between the mean and any number of standard deviations is also always constant. For example, the percentage of scores between the mean and two standard deviations either below or above the mean is always 47.72% in a normal distribution. The percentage of scores between the mean and three standard deviations either below or above the mean is always 49.87%. As our number of standard deviations increases, we approach in the area under the curve 50% of all scores on each side of the mean (50% below and above the mean would equal 100% of cases). But, no matter how many standard deviations away from the mean we go, we never include all 50% of cases that could occur on either side of the mean, because there is always the possibility of a very extreme score.

The number of standard deviations doesn't always come in ones, twos or threes. A score can be 1.45 or 2.73 standard deviations from the mean, or any number of standard deviations from the mean. The percentage of scores between the mean and 1.45 standard deviations on either side of the mean, for example, or 2.73 standard deviations on either side of the mean is also constant. The distribution table at the end of this chapter (called the *normal probability distribution table or* z *distribution table*) shows the percentage of scores between the mean and any specified number of standard deviations from 0.01 to 3.10 *either* below or above the mean. The number of standard deviations here is abbreviated "z" and is listed in the first column. The second column shows the percentage of scores between the mean and each specified number of standard deviations from the mean (42.65% of scores fall between the mean and 1.45 standard deviations, for example, or 49.68% of scores fall between the mean and 2.73 standard deviations in a normal distribution). The second column shows the percentage of scores for only one half of the normal curve, since both halves are symmetrical.

The percentage of scores between the mean and any raw score can be found by converting the raw score into a z score (also called a standard score). This z score tells you the distance from the mean of your raw score in standard deviations. To convert your raw score to the number of standard deviations it lies from the mean, you need to know the mean and standard deviation of the distribution the score comes from. Let us use as an example the score 600 from a distribution with a mean of 481.11 and a standard deviation of 116.69. To do the conversion:

1) Subtract the mean from the raw score. In this case, $600 - 481.11 = 118.89$.

2) Divide the result of step 1 by the standard deviation. This is your z value. If it has a negative sign, it tells you that the score is that number of standard deviations below the mean. If it has a positive sign, it tells you that the score is that number of standard deviations above the mean. In this case, $118.89/116.69 = 1.02$. This is the number of standard deviations above the mean of your raw score.

3) To find the percentage of scores which fall between the mean and the z value, use the normal distribution table at the end of this chapter. In this case, to find the percentage of scores which fall between the mean of 481.11 and 1.02 standard

deviations, look up 1.02 in the z column. The corresponding percentage of scores is 34.61%.

There are several things that we can do with this conversion technique:

Finding the Percentile Rank of a Raw Score

Situation 1: Your raw score is above the mean. The procedure is the following:

1) Subtract the mean from the raw score. In the above illustration, for example, $600 - 481.11 = 118.89$.

2) Divide the result of step 1 by the standard deviation. In the above case $118.89/116.69 = 1.02$, your z value.

3) Look up the z value you found in step 2 in the z distribution table to find the percentage of scores between the mean and that z value. In the above case, 34.61% of scores fall between the mean and a z of 1.02 or a raw score of 600.

4) Add 50% to the percentage from step 3. This is your percentile rank. In this case, $50\% + 34.61\% =$ a percentile rank of 84.61.

The following normal curve pictorially represents the position of the score you have been working with:

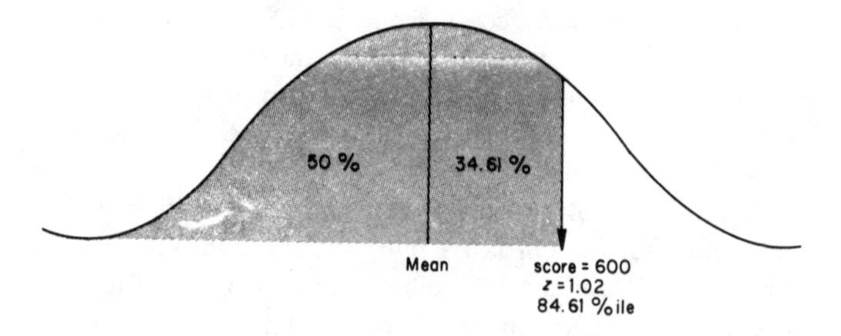

This is a quick method for finding the percentile rank of a score when you know the mean and standard deviation of the distribution that the score is from.

REMEMBER:

The PERCENTILE RANK of a score tells you the percentage of people who took the test who scored *at or below your score.*

Situation 2: Your raw score is below the mean. Suppose you want to find the percentile rank of a score of 400 from a distribution with a mean of 481.11 and a standard deviation of 116.69. The procedure is the following:

1) Subtract the mean from the raw score. In this case, $400 - 481.11 = -81.11$.

2) Divide the result of step 1 by the standard deviation. In this case, $-81.11/116.69 = -.70$, your z value. Remember that it has a negative sign because it indicates the raw score's number of standard deviations below the mean.

3) Look up the z value you found in step 2 in the z distribution table to find the percentage of scores between the mean and that z value. In this case, 25.80% of scores fall between the mean and a z of $-.70$ or a raw score of 400.

4) Subtract the percentage you found in step 3 from 50%. This is your percentile rank. In this case $50\% - 25.80\% = 24.20\%$.

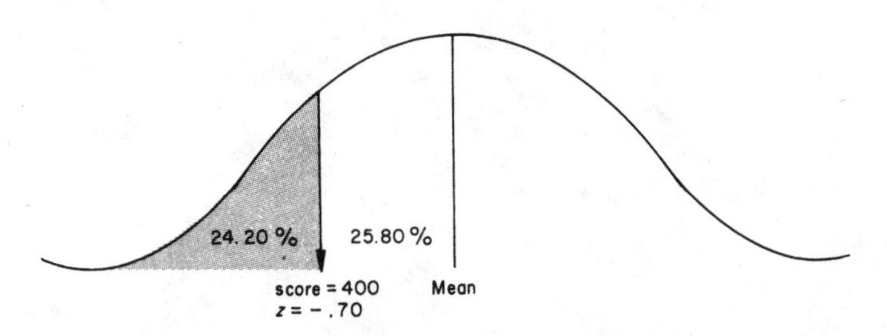

If your raw score is exactly the same as the mean, remember that your mean is equivalent to the 50th percentile.

Finding the Raw Score Corresponding to a Specified Percentile Rank

Situation 1: Your specified percentile rank is above the 50th percentile. Let's suppose, for example, using the same distribution with a mean of 481.11 and a standard deviation of 116.69, we want to find the score corresponding to the 90th percentile. We must do the following:

1) Subtract 50 from the specified percentile. In this case, $90 - 50 = 40$.

2) Find the z score, or number of standard deviations, in the normal distribution table corresponding to the percentage of cases computed in step 1. In this case, the z corresponding to approximately 40% (39.97% is the closest we can get in the table) is 1.28.

3) Multiply the result of step 2 by the standard deviation of the distribution of scores. In this case $(1.28) \times (116.69) = 149.36$.

4) Add the result of step 3 to the mean. In this case, $481.11 + 149.36 = 630.47$. This is the raw score corresponding to the 90th percentile.

The following normal curve pictorially represents the position of the score you have been working to find:

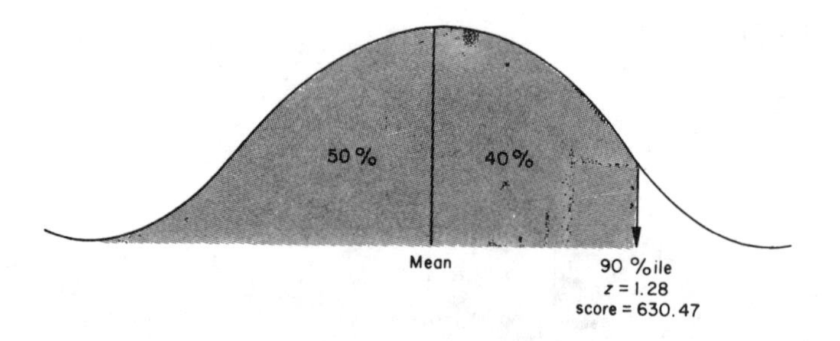

Situation 2: Your specified percentile rank is below the 50th percentile. Suppose, for example, you want to find the score at the 10th percentile from a distribution with a mean of 481.11 and a standard deviation of 116.69:

1) Subtract the percentile specified from 50%. In this case, $50 - 10 = 40\%$.

2) Find the z score, or number of standard deviations, in the normal distribution table, corresponding to the percentage calculated in step 1. In this case, the z corresponding to approximately 40% (39.97% is the closest we can get from the table) is 1.28.

3) Multiply the result of step 2 by the standard deviation of

the distribution of scores. In this case, $(1.28) \times (116.69) = 149.36$.

4) Subtract the result of step 3 from the mean. In this case, $481.11 - 149.36 = 331.75$. This is the raw score corresponding to the 10th percentile.

The following normal curve pictorially represents the position of the score you have been working to find:

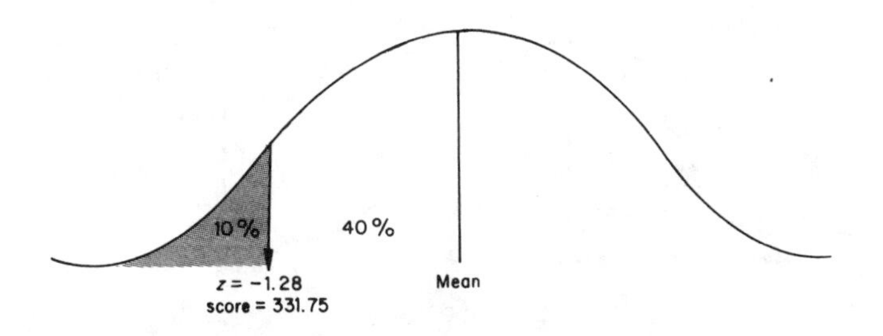

If your specified percentile rank is the 50th percentile, then all you need to do to find the score corresponding to it is to find the mean, which is the 50th percentile in a normal distribution.

Finding the Percentage of Scores Between Any Two Scores, Including the Mean

The shaded areas of the following curves represent the areas whose percentage of scores you are trying to find. Each curve represents a different situation. For each situation, the rules of procedure are at the right of the curve: Each procedure is based upon the three steps for finding the percentage of scores between the mean and any raw score outlined on p. 65.

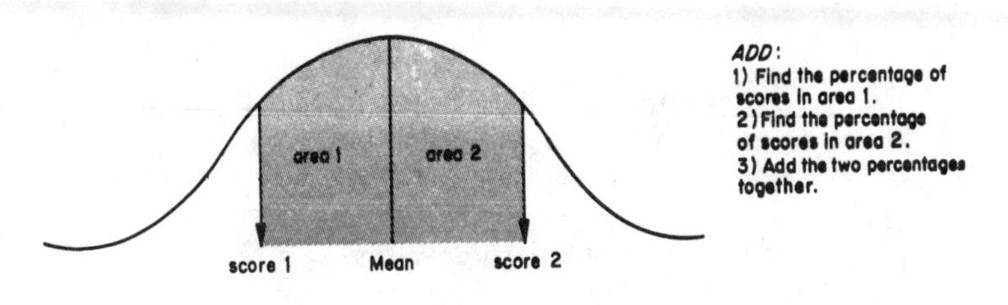

ADD:
1) Find the percentage of scores in area 1.
2) Find the percentage of scores in area 2.
3) Add the two percentages together.

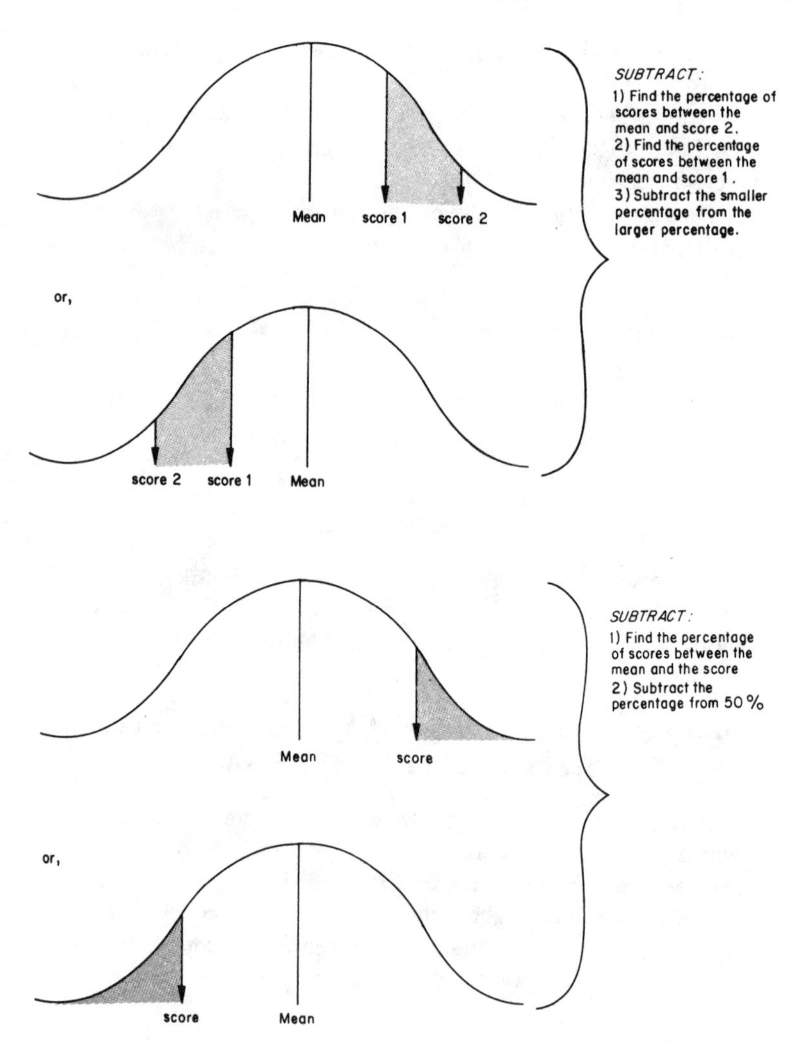

The procedures in this chapter are all based upon the following formula for creating standardized scores:

$$z = \frac{\text{score} - \text{mean}}{\text{standard deviation}}$$

Review Questions

1. What percentage of scores are below the mean in a normal distribution?
2. What does the z value tell you?
3. What does a z value with a negative sign mean?
4. What information do you need to convert a raw score into a standard score?
5. To find the percentage of scores between two raw scores, when do we add and when do we subtract?

Review Exercises

Exercise 1. Supervisors in corporations throughout the country were given an examination to test their knowledge of basic principles of group dynamics. The mean score was 50 with a standard deviation of 3.5. Assuming that scores were distributed normally, what percentage of the population would we expect to obtain scores of 58 or less?

Since the raw score is above the mean:

1) The raw score minus the mean = 58 -- 50 = 8.

2) The result of step 1 divided by the standard deviation = 8/3.5 = a z value of 2.29.

3) Using the z distribution table, 48.9% of scores fall between the mean and a z value of 2.29.

4) 50% + 48.9% = a percentile rank of 98.9.

The following normal curve pictorially represents the position of the score you have been working with:

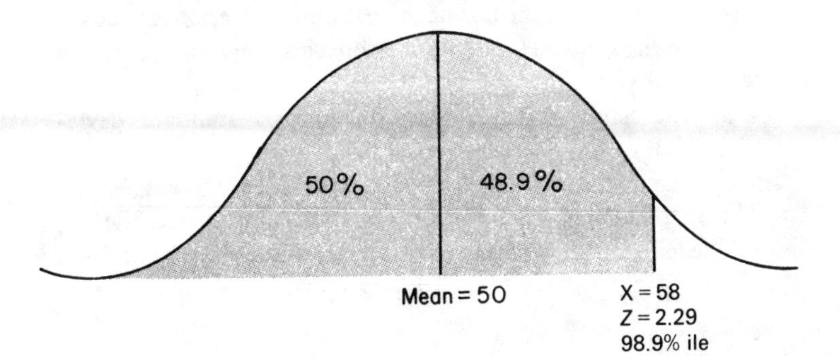

Exercise 2. What is the percentile rank of a score of 45 in the distribution in exercise 1 of this chapter?

Since the raw score is below the mean:

1) The raw score minus the mean = 45 – 50 = – 5.

2) The result of step 1 divided by the standard deviation = – 5/3.5 = a z value of – 1.43.

3) Using the z distribution table, 42.36% of scores fall between the mean and a z value of – 1.43.

4) 50% – 42.36% = a percentile rank of 7.64.

The following normal curve pictorially represents the position of the score you have been working with:

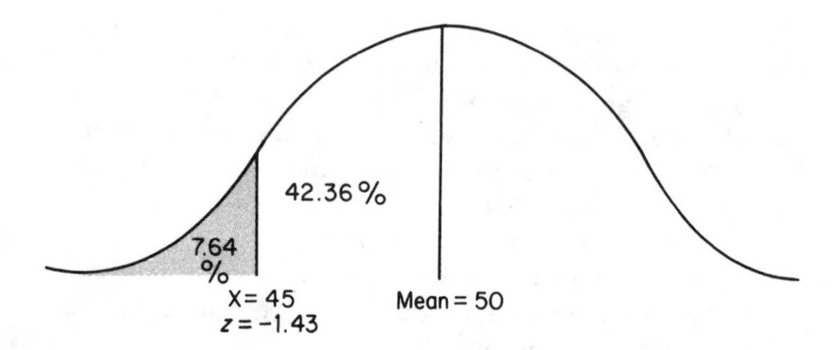

Exercise 3. What is the raw score corresponding to a percentile rank of 80% in the distribution in exercise 1 of this chapter?

Since the percentile rank is above the 50th percentile:

1) The specified percentile minus 50% = 80% – 50% = 30%.

2) The z score in the normal distribution table corresponding to approximately 30% (29.95% is the closest we can get in the table) is .84.

3) The result of step 2 multiplied by the standard deviation = (.84) × (3.5) = 2.94.

4) The sum of 2.94 plus the mean = 2.94 + 50 = 52.94. This is the raw score corresponding to the 80th percentile.

The following normal curve pictorially represents the position of the score you have been working to find:

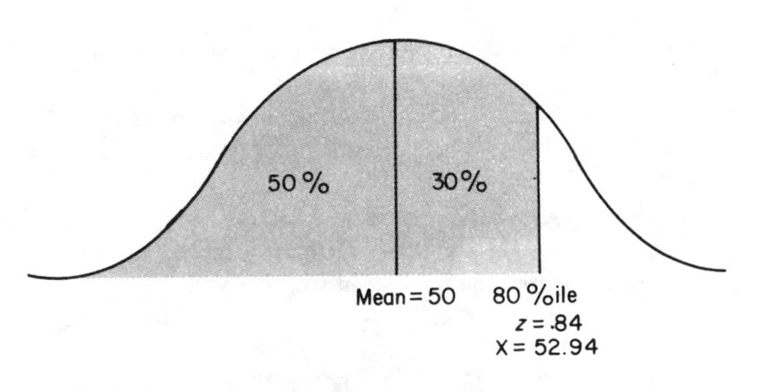

Exercise 4. What is the raw score corresponding to a percentile rank of 40 in the distribution in exercise 1 of this chapter?

Since the percentile rank is below the 50th percentile:

1) 50% minus the specified percentile = 50% – 40% = 10%.

2) The z score in the distribution table corresponding to approximately 10% (9.87% is the closest we can get in the table) is .25.

3) The result of step 2 multiplied by the standard deviation = (.25) × (3.5) = .88.

4) The mean minus the result of step 3 = 50 – .88 = 49.12.

The following normal curve pictorially represents the position of the score you have been working to find:

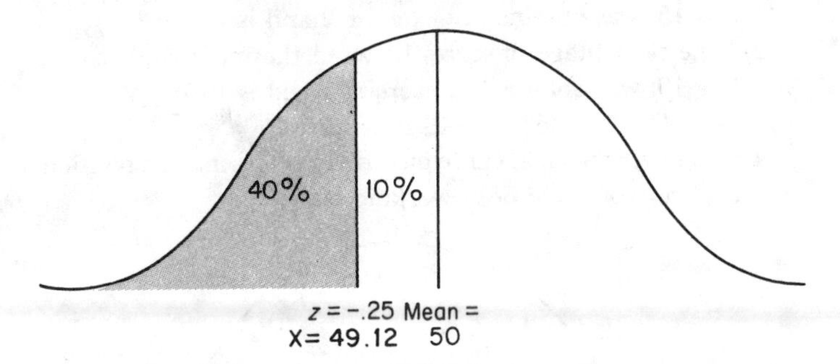

Exercise 5. What is the percentage of scores between 40 and 60 in the distribution in exercise 1 of this chapter?

Since the scores are on both sides of the mean:

1) Area 1: ($score_1$ – mean)/standard deviation = (40 – 50)/

$3.5 = -10/3.5 = -2.86$. The percentage of scores in area $1 = 49.79\%$.

2) Area 2: $(\text{score}_2 - \text{mean})/\text{standard deviation} = (60 - 50)/3.5 = 10/3.5 = 2.86$. The percentage of scores in area $2 = 49.79\%$.

3) $49.79\% + 49.79\% = 99.58\%$ of scores.

The following normal curve pictorially represents the positions of the scores you have been working with:

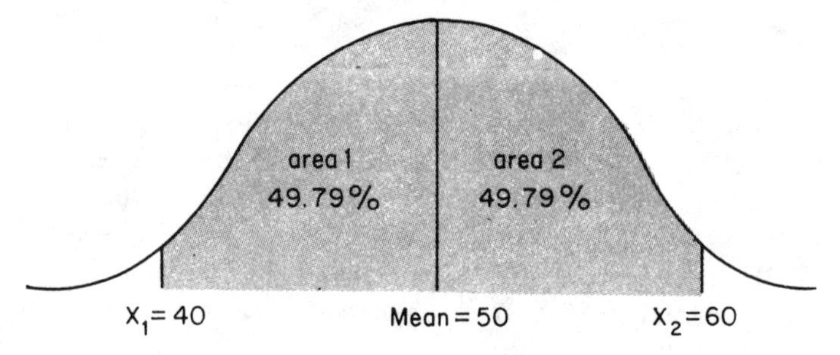

Exercise 6. What is the percentage of scores between 40 and 45 in the distribution in exercise 1 of this chapter?

Since the scores are on one side of the mean but the area they include does not contain a tail of the distribution:

1) The percentage of scores between the mean and score_2, which is 45, was obtained in exercise 2 and is 42.36%.

2) The percentage of scores between the mean and score_1, which is 40, was obtained in exercise 5 and is 49.79%.

3) $49.79\% - 42.36\% = 7.43\%$ of scores.

The following normal curve pictorially represents the position of the scores you have been working with:

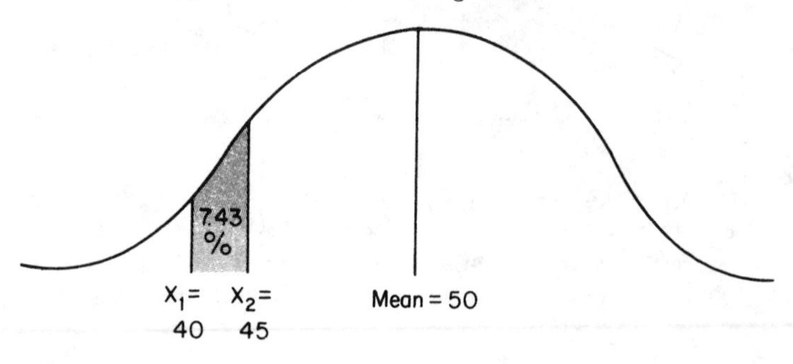

Exercise 7. What is the percentage of scores above 55 in the distribution in exercise 1 of this chapter?

Since the area we are finding includes the tail of the normal curve:

1) (score – mean)/standard deviation = (55 – 50)/3.5 = 5/3.5 = 1.43.

The percentage of scores between the mean and a score of 55 = 42.36%.

2) 50% – 42.36% = 7.64% of scores.

The following normal curve pictorially represents the position of the score you have been working with:

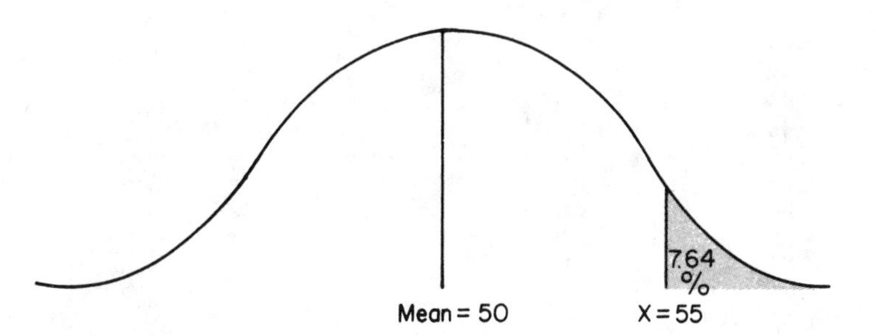

Table 6.1 Normal Probability Distribution Table

z value	area between mean and *z* value (%)	*z* value	area between mean and *z* value (%)
0.00	0.00	0.15	5.96
0.01	0.40	0.16	6.36
0.02	0.80	0.17	6.75
0.03	1.20	0.18	7.14
0.04	1.60	0.19	7.53
0.05	1.99	0.20	7.93
0.06	2.39	0.21	8.32
0.07	2.79	0.22	8.71
0.08	3.19	0.23	9.10
0.09	3.59	0.24	9.48
0.10	3.98	0.25	9.87
0.11	4.38	0.26	10.26
0.12	4.78	0.27	10.64
0.13	5.17	0.28	11.03
0.14	5.57	0.29	11.41

Table 6.1 *(Cont.)*

z value	area between mean and *z* value (%)	*z* value	area between mean and *z* value (%)
0.30	11.79	0.75	27.34
0.31	12.17	0.76	27.64
0.32	12.55	0.77	27.94
0.33	12.93	0.78	28.23
0.34	13.31	0.79	28.52
0.35	13.68	0.80	28.81
0.36	14.06	0.81	29.10
0.37	14.43	0.82	29.39
0.38	14.80	0.83	29.67
0.39	15.17	0.84	29.95
0.40	15.54	0.85	30.23
0.41	15.91	0.86	30.51
0.42	16.28	0.87	30.78
0.43	16.64	0.88	31.06
0.44	17.00	0.89	31.33
0.45	17.36	0.90	31.59
0.46	17.72	0.91	31.86
0.47	18.08	0.92	32.12
0.48	18.44	0.93	32.38
0.49	18.79	0.94	32.64
0.50	19.15	0.95	32.89
0.51	19.50	0.96	33.15
0.52	19.85	0.97	33.40
0.53	20.19	0.98	33.65
0.54	20.54	0.99	33.89
0.55	20.88	1.00	34.13
0.56	21.23	1.01	34.38
0.57	21.57	1.02	34.61
0.58	21.90	1.03	34.85
0.59	22.24	1.04	35.08
0.60	22.57	1.05	35.31
0.61	22.91	1.06	35.54
0.62	23.24	1.07	35.77
0.63	23.57	1.08	35.99
0.64	23.89	1.09	36.21
0.65	24.22	1.10	36.43
0.66	24.54	1.11	36.65
0.67	24.86	1.12	36.86
0.68	25.17	1.13	37.08
0.69	25.49	1.14	37.29
0.70	25.80	1.15	37.49
0.71	26.11	1.16	37.70
0.72	26.42	1.17	37.90
0.73	26.73	1.18	38.10
0.74	27.04	1.19	38.30

Table 6.1 (*Cont.*)

z value	area between mean and z value (%)	z value	area between mean and z value (%)
1.20	38.49	1.65	45.05
1.21	38.69	1.66	45.15
1.22	38.88	1.67	45.25
1.23	39.07	1.68	45.35
1.24	39.25	1.69	45.45
1.25	39.44	1.70	45.54
1.26	39.62	1.71	45.64
1.27	39.80	1.72	45.73
1.28	39.97	1.73	45.82
1.29	40.15	1.74	45.91
1.30	40.32	1.75	45.99
1.31	40.49	1.76	46.08
1.32	40.66	1.77	46.16
1.33	40.82	1.78	46.25
1.34	40.99	1.79	46.33
1.35	41.15	1.80	46.41
1.36	41.31	1.81	46.49
1.37	41.47	1.82	46.56
1.38	41.62	1.83	46.64
1.39	41.77	1.84	46.71
1.40	41.92	1.85	46.78
1.41	42.07	1.86	46.86
1.42	42.22	1.87	46.93
1.43	42.36	1.88	46.99
1.44	42.51	1.89	47.06
1.45	42.65	1.90	47.13
1.46	42.79	1.91	47.19
1.47	42.92	1.92	47.26
1.48	43.06	1.93	47.32
1.49	43.19	1.94	47.38
1.50	43.32	1.95	47.44
1.51	43.45	1.96	47.50
1.52	43.57	1.97	47.56
1.53	43.70	1.98	47.61
1.54	43.82	1.99	47.67
1.55	43.94	2.00	47.72
1.56	44.06	2.01	47.78
1.57	44.18	2.02	47.83
1.58	44.29	2.03	47.88
1.59	44.41	2.04	47.93
1.60	44.52	2.05	47.98
1.61	44.63	2.06	48.03
1.62	44.74	2.07	48.08
1.63	44.84	2.08	48.12
1.64	44.95	2.09	48.17

Table 6.1 (*Cont.*)

z value	area between mean and *z* value (%)	*z* value	area between mean and *z* value (%)
2.10	48.21	2.55	49.46
2.11	48.26	2.56	49.48
2.12	48.30	2.57	49.49
2.13	48.34	2.58	49.51
2.14	48.38	2.59	49.52
2.15	48.42	2.60	49.53
2.16	48.46	2.61	49.55
2.17	48.50	2.62	49.56
2.18	48.54	2.63	49.57
2.19	48.57	2.64	49.59
2.20	48.61	2.65	49.60
2.21	48.64	2.66	49.61
2.22	48.68	2.67	49.62
2.23	48.71	2.68	49.63
2.24	48.75	2.69	49.64
2.25	48.78	2.70	49.65
2.26	48.81	2.71	49.66
2.27	48.84	2.72	49.67
2.28	48.87	2.73	49.68
2.29	48.90	2.74	49.69
2.30	48.93	2.75	49.70
2.31	48.96	2.76	49.71
2.32	48.98	2.77	49.72
2.33	49.01	2.78	49.73
2.34	49.04	2.79	49.74
2.35	49.06	2.80	49.74
2.36	49.09	2.81	49.75
2.37	49.11	2.82	49.76
2.38	49.13	2.83	49.77
2.39	49.16	2.84	49.77
2.40	49.18	2.85	49.78
2.41	49.20	2.86	49.79
2.42	49.22	2.87	49.79
2.43	49.25	2.88	49.80
2.44	49.27	2.89	49.81
2.45	49.29	2.90	49.81
2.46	49.31	2.91	49.82
2.47	49.32	2.92	49.82
2.48	49.34	2.93	49.83
2.49	49.36	2.94	49.84
2.50	49.38	2.95	49.84
2.51	49.40	2.96	49.85
2.52	49.41	2.97	49.85
2.53	49.43	2.98	49.86
2.54	49.45	2.99	49.86

Table 6.1 (*Cont.*)

z value	area between mean and *z* value (%)	*z* value	area between mean and *z* value (%)
3.00	49.87	3.06	49.89
3.01	49.87	3.07	49.89
3.02	49.87	3.08	49.90
3.03	49.88	3.09	49.90
3.04	49.88	3.10	49.90
3.05	49.89		

Source: Adapted from Table I of Hald, *Statistical Tables and Formulas*, published by John Wiley and Sons, Inc., New York. Used by permission of the author and publishers.

7

Probability and the Binomial

Probability is important in statistics because hypothesis testing is based upon probability. A hypothesis is a prediction about the result you will find when you collect information and analyze it. When looking at information, it is important to be able to decide whether you got the results that you did because of chance variation or because some event other than chance affected your results. Probability helps us to figure out the likelihood that the result obtained occurred by chance alone. If the results we obtained occurred by chance alone, they are often not as interesting to us as they are when something other than chance is occurring.* If something other than chance is occurring, we have a *statistically significant* difference (e.g., between groups) or relationship (between variables) and we can choose between two conflicting hypotheses at a certain level of probability of having made a correct choice.

Probabilities are usually expressed in research as (1) a proportion, or (2) a percentage. When expressed as a proportion, they vary on a scale from 0 to 1.00 and are usually expressed as a decimal (e.g., .74, .50, etc.). When expressed as a percentage, the proportion is multiplied by 100 and probabilities vary on a scale from 0 to 100 (e.g., 74%, 50%, etc.). Probabilities

*An exception would be when we gathered evidence to support the conclusion that another researcher's significant results were not apparent in our research.

can also be expressed as the number of chances in 100 or as odds (e.g., 3 to 1).

Probability of a Single Event

To figure out the probability of a single event occurring, the formula is

$$\text{probability of event} = \frac{\text{number of times event occurred}}{\text{total number of times event could have occurred}}$$

If there are ten people in a room and three have college degrees, the probability of picking someone who has a college degree if all the names are put on slips of paper and tossed into a hat is .30. In other words, college degrees occurred 3 times. However, since there are ten people, college degrees could have occurred 10 times. Therefore, $3/10 = .30$.

This principle of probability, as well as the others that follow, can be applied to areas under the normal curve. For example, in a distribution where the mean is 481.11 and the standard deviation is 116.69, a raw score of 600 falls at the 84.61 percentile (see p. 66 for the calculations). Therefore, the probability of selecting a student from those who took the exam who scored at or below 600 is .8461.

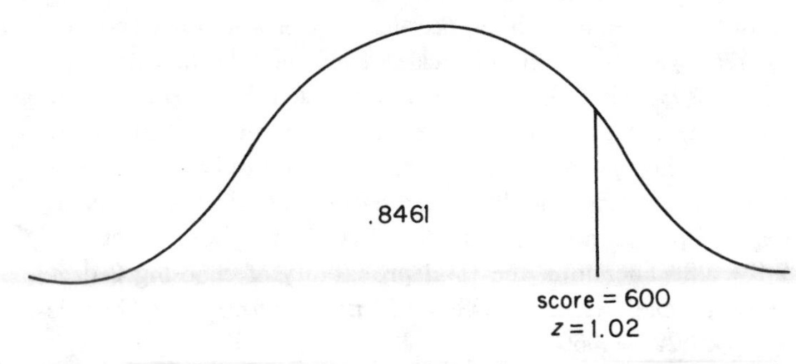

The probability of selecting a student who scores at or below 600 added to the probability of choosing a student who scores above 600 is always 1.00 (100 if probability is expressed as percentages). In other words, the probability of an event occurring *plus* the probability of an event *not* occurring equals 1.00 (100 if probability is expressed as percentages).

Probability of Two or More Events

Situation 1: *Multiplication rule*—the probability of *both* or *all* events occurring.

If we want to determine the probability of joint or successive occurrence of two or more events, we must *multiply* the separate probabilities of each event. Express probability as a proportion when multiplying.

Independent events. If each event is *independent*, that is, if the probability of each event occurring is unaffected by whether or not any other event occurred, we multiply the separate probabilities of each. For example, in the distribution referred to above, the probability of selecting a student who scored at or below 600 is .8461. If that student's name were put back into the population pool (called *sampling with replacement*), the probability of selecting two students who scored at or below 600 would be $(.8461) \times (.8461) = .7159$.

Conditional probability. If the probability of one event occurring is *dependent* upon whether or not the other event occurred, we multiply the probability of the first event by the probability of successive events *given that the previous event has occurred.* In the example of a room with ten people, three of whom have college degrees, the probability of picking someone who has a college degree is .30 when the first person is selected. If someone with a college degree is selected and their name is *not* put back in the hat (called *sampling without replacement*), there are only two degreed people left out of nine, so the probability of selecting a second degreed person now that the first has already been selected is $2/9 = .22$. Therefore, the total probability of choosing 2 degreed people when drawing 2 names from the hat is $(.30) \times (.22) = .066$.

Situation 2: *Addition rule*—the probability of *either* of several events occurring.

If we want to determine the probability of occurrence of either of several events, we must *add* the separate probabilities of each event.

Mutually exclusive events. If both events cannot occur simultaneously, we add the separate probabilities of each. For example, suppose we have a distribution in which the probability of selecting a person with an hourly wage between $9 and $10 is .3413. Suppose also that the probability of selecting a person with an hourly wage of $12 or more is .0228. The probability of selecting *either* a person with an hourly wage between $9 and $10 *or* a person with an hourly wage of $12 or more would be $.3413 + .0228 = .3641$.

Suppose also that there are several possible combinations of ways events can occur that ordinarily make use of the multiplication rule. For example, suppose that we want to know the probability of selecting first a person with an hourly wage between $9 and $10 and then another person with an hourly wage of $12 or more. Ordinarily we would multiply the probabilities of each event: $(.3413) \times (.0228) = .0078$. However, if we were not concerned with the order of selection, there would be 2 ways of obtaining a person with an hourly wage between $9 and $10 and a person with an hourly wage of $12 or more. We can select them in that order, or we could select the person with an hourly wage of $12 first and the lesser paid worker second. In situations with several possible combinations that satisfy the criteria, we must combine the multiplication and addition rules and add the probabilities, obtained through multiplication, of all the possible ways the combination of events could occur. In this case, $.0078 + .0078 = .0156$.

Non-mutually exclusive events. If events can occur simultaneously, we add the separate probabilities of each and subtract the probability of their joint occurrence. Suppose we have a distribution where the probability of earning an hourly wage between $9 and $10 is .3413. Suppose the probability of being female is .25 and that the probability of being female and earning between $9 and $10 is .17. To obtain the probability of selecting someone who either earns between $9 and $10 or is female, $.3413 + .25 - .17 = .4213$.

The Binomial Distribution

Notions of probability can be applied to different models of sampling distributions. In social science, the *normal curve* and the *binomial distribution* are models which are frequently used. Much hypothesis testing is based upon the normal curve, as you will see in subsequent chapters. However, included here is an illustration of the application of the binomial. The binomial is applicable when each case results in either one of two mutually exclusive outcomes which have a probability of 1.0 when added together (e.g., head and tail when tossing a coin).

The binomial distribution approaches the normal distribution in shape when the number of cases is large. If the number of cases is small, the binomial still approaches the normal distribution in shape if the probabilities of the two possible outcomes in the binomial each approach .50.

Suppose that a candidate named Scott is running for office and we want to get in advance of the vote some idea of whether or not Scott is the favored candidate. We take a sample of 100 registered voters. We want to know the possible range of the number of registered voters sampled who will say they favor Scott, if, in fact, Scott and her opponent are equally favored.

If Scott and her opponent are equally favored, every time we take a sample of registered voters, we will not get 50% who say they will vote for Scott and 50% who say they will vote for her opponent. There is variability in repeated sampling and a sample may result, for example, in 55 out of 100 saying they favor Scott. The binomial helps us determine the range of variability that is probable if the candidates *are* equally favored.

To determine the range of variability of the sampled voters, do the following:

1) Multiply the number in the sample by the probability that voters will vote for Scott if both candidates are equally favored. There are 100 cases in the sample. If both candidates are equally favored, the probability that voters will vote for Scott is .50. Thus, $(100) \times (.50) = 50$. This is the mean of your binomial.

2) Multiply the result of step 1 by the probability that voters will *not* vote for Scott. Remember that both outcomes have a probability of 1.0 when added together. In this case, the

probability that voters *will* vote for Scott is .50. Therefore the probability that voters will *not* vote for Scott is also .50 (.50 + .50 = 1.00). In this case, (50) × (.50) = 25.

3) Compute the square root of the result of step 2. This is the standard deviation of your binomial. In this case, the square root of 25 is 5.

4) Decide whether you want to know the range of variability of 95% or 99% of the sampled voters. If 95%, multiply the result of step 3 by 1.96. If 99%, multiply the result of step 3 by 2.58. In this case, let us find the range of variability of 95% of samples of voters. (5) × (1.96) = 9.8.

5) Subtract the result of step 4 from the result of step 1. This is the lower end of the range. In this case, 50 – 9.8 = 40.2.

6) Add the result of step 4 to the result of step 1. This is the upper end of the range. In this case, 50 + 9.8 = 59.8. Therefore, 95% of the samples taken with repeated sampling will result in between 40.2 and 59.8 voters saying they favor Scott, if the candidates are in fact equally favored. To be able to conclude that Scott is the favored candidate we would need to have more than 59.8 voters say they favored Scott in our sample of 100. Since this would be an improbable event if the candidates were equally favored, we could conclude either that this was a rare event or that Scott was favored. In statistics, we would conclude that the latter was probably so.

The upper and lower ends of the range of variability of 95% of samples of voters in this example are what are known as *confidence intervals*. Often we don't have hypotheses about exact values, but are interested in discovering the range of values within which a hypothesis is either true or false. When looking at the range of variability of 95% of values, we are looking for the 5% confidence interval. When looking at the range of variability of 99% of values, we are establishing a 1% confidence interval.

Review Questions

1. What does a statistically significant difference mean?
2. How do you determine the probability of a single event occurring?

3. When is the multiplication rule used?
4. What does conditional probability mean?
5. When is the addition rule used?
6. What does mutually exclusive events mean?
7. When is the binomial applicable?
8. What is meant by confidence intervals?

Review Exercises

Exercise 1. In a local business there are 1,400 female workers compared to 800 male workers. What is the probability (p) of randomly selecting a female in a single draw for a sample?

$$1400 + 800 = 2200 \text{ workers}$$

$$p = 1400/2200 = .6364 \text{ or } 63.64\%$$

Exercise 2. In the normal distribution in the review exercises for Chapter 6, what is the probability of randomly selecting a supervisor who obtained a score of 58 or less in a single draw for a sample?

Since 98.9% of the population obtained a score of 58 or less, the probability would be 98.9%.

Exercise 3. If the probability of randomly selecting a supervisor who obtained a score below 45 in the normal distribution in the review exercises for Chapter 6 is .0764, what is the probability of randomly selecting two supervisors who obtained a score below 45 in two draws for a sample? Assume replacement.

We use the multiplication rule for independent events:

$$(.0764) \times (.0764) = .0058 \text{ or } .58\%$$

Exercise 4. What is the probability of randomly selecting two females in two draws for a sample from a group consisting of 40 females and 60 males? Assume no replacement.

We use the multiplication rule for dependent events (conditional probability) since there is no replacement.

p of drawing the first female = 40/100 = .4000.

p of drawing the second female = 39/99 = .3939.

$(.4000) \times (.3939) = .1576$ or 15.76%

Exercise 5. In chapter 6 review exercises, what is the probability of randomly selecting either a supervisor who obtained a score below 45 or a supervisor who obtained a score above 55 in a single draw for a sample from the normal distribution?

We use the addition rule for mutually exclusive events:

$$.0764 + .0764 = .1528 \text{ or } 15.28\%$$

Exercise 6. What is the probability of randomly selecting a supervisor who obtained a score below 45 and a supervisor who obtained a score above 60 in two draws for a sample from the normal distribution in the review exercises in chapter 6?

Since the events can occur in two distinct orders:

a) p of selecting first a supervisor scoring below 45 and second a supervisor scoring above 60 = .0764 × .0021 = .0002 or 0.02%.

b) p of selecting first a supervisor scoring above 60 and second a supervisor scoring below 45 = .0021 × .0764 = .0002 or 0.02%.

total probability = .0004 or 0.04%

Exercise 7. A journalist wishes to know if the proportion of adults in New York City who exercise regularly is the same as the national proportion of .40. A sample of 50 New Yorkers is drawn. The proportion of the sample who exercise regularly is .30. What should the journalist conclude?

1) The number in the sample multiplied by the probability that New Yorkers exercise regularly if in fact they exercise as much as the national proportion = (50) × (.40) = 20.

2) The result of step 1 multiplied by the probability that New Yorkers will *not* exercise regularly, if in fact they exercise as much as the national proportion = (20) × (.60) = 12.

3) The square root of 12 is 3.46.

4) If we are interested in the range of variability of 95% of the sampled New Yorkers, (3.46) × (1.96) = 6.78.

5) The result of step 1 minus the result of step 4 = 20 − 6.78 = 13.22, the lower end of the range.

6) The sum of the result of step 1 and the result of step 4 = 20 + 6.78 = 26.78, the upper end of the range.

Therefore, 95% of the samples taken with repeated sampling will result in between 13.22 and 26.78 New Yorkers saying they exercise regularly if, in fact, the proportion who do so is equal to the national proportion. In this sample, .30 of 50 New Yorkers or 15 New Yorkers said they exercise regularly. That number falls between 13.22 and 26.78. Therefore, it is not an improbable sample result if, in fact, the proportion of New Yorkers who exercise regularly and the national proportion are equal.

8

One-sample z and t Tests

How to Compare the Mean of One Sample to the Mean of a Population

We can apply the concept of probability to the normal curve in several ways:

1) When we convert raw scores on a normally distributed variable into z scores, we are transforming those raw scores into units of the normal curve. The total area in a normal distribution equals 1.00. We can express any portion of the normal curve as a proportion of the total area under the curve.

The *tails* (also called *critical regions*) are the extreme ends of the normal curve. A score has only a *small probability* of falling into a tail region:

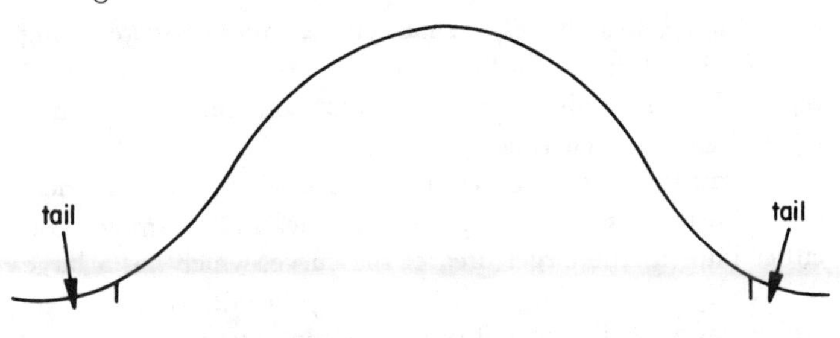

The exact probability of a score falling into a tail region equals the proportion of area under the normal curve that is alloted to the tails. For example, if the probability of a score falling into a tail region is 0.05, this means that 0.05 of the curve is marked off as being in the tails. By convention, the proportion of area generally allotted to the tails is 0.05, 0.01 or 0.001. We can choose *any* of these. We can put the whole proportion of area into one of the tails (either the one on the left side or the one on the right side of the curve) or divide it in half and put half in each of the two tails. 0.05 of the area, for example, would become 0.025 in each tail ($0.025 + 0.025 = 0.05$) if we wanted to use both tails of the distribution:

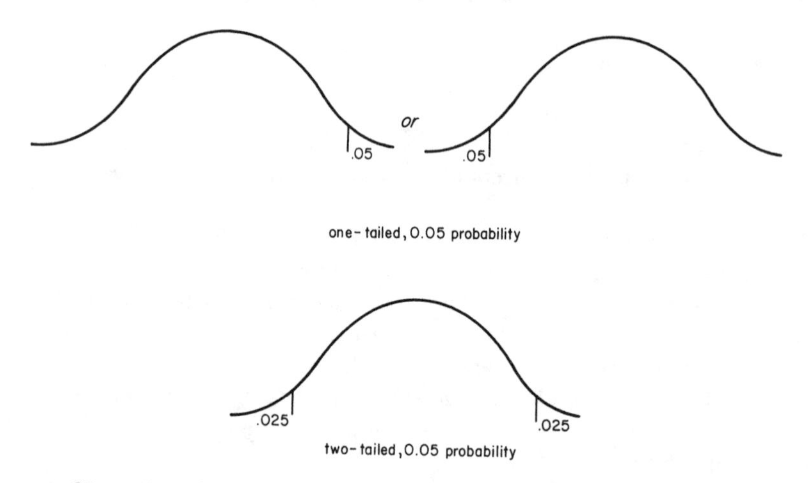

Since there is usually only a small probability of a score falling into a tail, when it does fall into a tail, that is, into a critical region, it is grounds for assuming that something other than mere chance is occurring.

3) Statistical tests are designed to see whether a numerical value (in the case of this chapter, the mean of a sample) falls either into (a) the larger area of the curve, which has a large probability of occurring by chance, or (b) the tail(s) of the curve, which has a small probability of occurring by chance.

4) Hypothesis testing is the setting up of two contradictory hypotheses: (a) the *null hypothesis* which predicts that the numerical value falls into the larger area of the curve and is therefore probably part of that distribution, and (b) the *research*

or alternative hypothesis which predicts that the numerical value falls into the tail(s) and is therefore probably not part of the distribution represented by the curve.

5) The research or alternative hypothesis sets up either a one-tailed or two-tailed test. If the hypothesis predicts, for example, that the sample mean is different from the population mean, that is a two-tailed test because there is no direction indicated in the prediction. Therefore, the sample mean may fall into a tail on either side of the distribution. Words like "difference" or "relationship" are generally used in two-tailed tests.

If the hypothesis predicts, for example, that the sample mean is greater or less than the population mean, that is a one-tailed test because direction is indicated in the prediction. Therefore, you are specifying which end of the distribution will have a tail, or critical region. Words like "higher", "lower", "greater", "fewer" are generally used in one-tailed tests. If it is possible to predict a direction for the tail, it is generally better to do so, because then the proportion of area under the curve allocated for the tail(s) does not have to be divided in half to construct two critical regions. Instead, a single critical region will be twice as large. Therefore, there is twice as much probability of rejecting the null hypothesis and accepting the research hypothesis, which is often what researchers will prefer to do.

6) Since hypothesis testing is based upon probability, sometimes we accept the wrong hypothesis. There are two types of errors we can make this way in hypothesis testing: *type I error* and *type II error*.

A type I error (also called *alpha*) occurs when the null hypothesis is rejected in favor of the research hypothesis, even though the null hypothesis is the one that is really true. The probability of making a type I error is the size of the critical region or regions. For example, if 0.05 of the normal curve area is alloted to the tails, the probability of making a type I error is 5%.

A type II error (also called *beta*) occurs when the null hypothesis is *not* rejected in favor of the research hypothesis, even though the null hypothesis is really false. There is a greater probability of making a type II error than a type I error. The smaller the critical region (e.g., 0.01 or 0.001 instead of 0.05

probability), the less the probability of a type I error and the greater the probability of a type II error. The more conservative you wish to be in choosing a research hypothesis, the smaller your critical region should be.

We can never tell with any particular research project whether a type I or type II error has been made because our inferences are based upon probability rather than fact. This is one of the reasons why it is desirable for the work of one researcher or research team to be repeated by others. If others obtain results similar to those of the original research, there is less chance that a type I error is being made.

Comparing a Sample Mean to a Population Mean

If you were to take random samples of the same size from a population and compute the mean of each sample, the distribution of those means would form a normal curve:

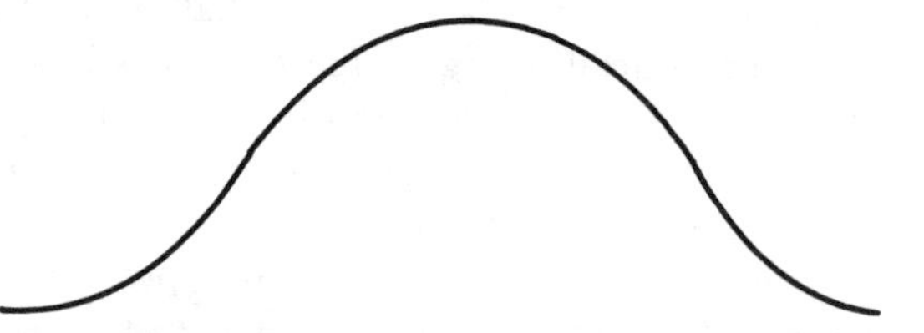

The standard deviation of the distribution of means would be called the *standard error*. Therefore, if we have a particular sample mean, we can use probability to determine if the mean of any particular sample is likely to come from that population's distribution of sample means or if the mean of that sample is significantly different and therefore probably comes from another population's distribution. To do this, we need to know (1) the mean of the sample and of the population; (2) the standard deviation of the population, or of the sample, and (3) the number of cases in the sample. It is assumed that the sample has been taken from a normally distributed population.

Situation 1: The z test—the population standard deviation is known.

Let us examine a situation in which we have a randomly selected sample of 50 cases. This is a sample of assembly-line workers who have participated in a motivational program designed to increase productivity. Their mean productivity is the assembly of 14.31 parts per hour with a standard deviation of 2.18. The mean productivity in general of assembly-line workers in this factory is the assembly of 13.49 parts per hour with a standard deviation of 1.97. We want to know if the motivational program increases productivity. Other ways of saying this are (1) *two-tailed test*: we want to know if there is a significant *difference* in the mean productivity of workers who participated in the motivational program and the general population of assembly-line workers (i.e., does the sample come from a population with a different mean?) or (2) *one-tailed test*: we want to know if workers who participated in the motivational program have significantly *higher* mean productivity than the general population of assembly-line workers.

Obviously, in just looking at the figures in this example, we can see that mean productivity of the motivational group is slightly higher. However, we need to test to see if that is probably merely an accident of sampling *or* probably a real difference in productivity.

To set up a test of a hypothesis, we go through the following steps:

1) Construct a *null hypothesis*. The null hypothesis here should state that the sample mean comes from the population's distribution of sample means. In symbolic form, it looks like this: $\mu = \mu_0$.

2) Construct an *alternative* or *research hypothesis*. The alternative or research hypothesis should state either that (a) the sample mean comes from a *different* population. This is a *two-tailed* test. In symbolic form, the hypothesis looks like this: $\mu \neq \mu_0$. Or (b) the sample mean comes form a population with a mean *greater than* (or *less than*) the population mean described here. This is a *one-tailed* test. In symbolic form, it looks like this: $\mu > \mu_0$ for "greater than" (or $\mu < \mu_0$ for "less than"). In this case, let us choose a one-tailed hypothesis: $\mu > \mu_0$.

3) Choose a *test statistic*. Select the one-sample z test because

we have a sample from a normally distributed population and because we know the number of cases in the sample, the sample mean, population mean and, particularly, the population standard deviation. We also know the sample standard deviation in this example, but it is extra information. We do not need to use it because it is better to use the population standard deviation if we know it.

4) Choose a *significance level*. We must choose either 0.05, 0.01 or 0.001 for the tail(s). The larger your significance level (0.05 is the largest), the greater your chance of rejecting your null hypothesis. Let us select 0.01 for this example.

5) Select a *sampling distribution*. With z tests, the sampling distribution is always the normal probability curve. You can see the normal curve distribution table in chapter 6 (pp. 75–79).

6) Select a *critical value*. We must select a critical value with which we will compare our *calculated value*, obtained using a version of the z formula. If our calculated value is equal to or greater than our critical value, we reject our null hypothesis and accept our research hypothesis. We find our critical value in the normal z distribution table. They are the values beyond which a z value will fall into the tail or critical region. The following are the critical z values for various significance levels with both one-tailed and two-tailed hypotheses. These values have been taken from the normal curve distribution table.

One-tailed Values

significance level	critical z value
0.05	1.65
0.01	2.33
0.001	3.08

Two-tailed Values

significance level	critical z value
0.05	1.96
0.01	2.58
0.001	3.30

For a one-tailed test at the 0.01 level, therefore, our critical z value is 2.33.

7) We now *carry our test.* The main purpose of the one-sample z test is to calculate if the sample mean comes from a population that is different from the population represented in the problem.

To carry out a one-sample z test, do the following:

a) Subtract the population mean from the sample mean. In this case, $14.31 - 13.49 = 0.82$.

b) Compute the square root of the number of cases in the sample. The square root of 50 is 7.07.

c) Divide the population standard deviation by the result of step 7b. In this case, $1.97/7.07 = .279$. If you *do not know* the *population* standard deviation, use the t test on p. 96 instead.

d) Divide the result of step 7a by the result of step 7c. This is your calculated z value.* In this case, $0.82/.279 = 2.94$.

These steps are based upon the following formula for calculating a z value when comparing a sample mean to a population mean:

$$z = \frac{\text{sample mean} - \text{population mean}}{\text{pop. standard deviation} / \sqrt{\text{number of cases in sample}}}$$

or, in symbolic form,

$$z = \frac{\overline{X} - \mu}{\sigma / \sqrt{N}} = \frac{14.31 - 13.49}{1.97 / \sqrt{50}} = 2.94$$

There are some familiar symbols here and some new ones:
$\overline{X}$ is the symbol for sample mean.
μ is the symbol for population mean.
σ is the symbol for population standard deviation.
$\sigma/\sqrt{N}$ is the symbol for standard error.
N is the symbol for number of cases.

8) We compare 2.94 (our calculated value) to the critical value (the distribution table value) of 2.33. Since our calculated

*If your calculated value has a negative sign, do not worry. This happens whenever your sample mean is less than your population mean. This is only a problem if you have set up a one-tailed test which predicted that your sample mean would be significantly greater than your population mean and have therefore set up a critical region on the wrong end of the normal curve.

value is higher than the critical value, the decision we make is to reject our null hypothesis and accept our research hypothesis. We conclude that mean productivity of assembly-line workers exposed to the motivational program is significantly higher than mean productivity in general of assembly-line workers in the factory.

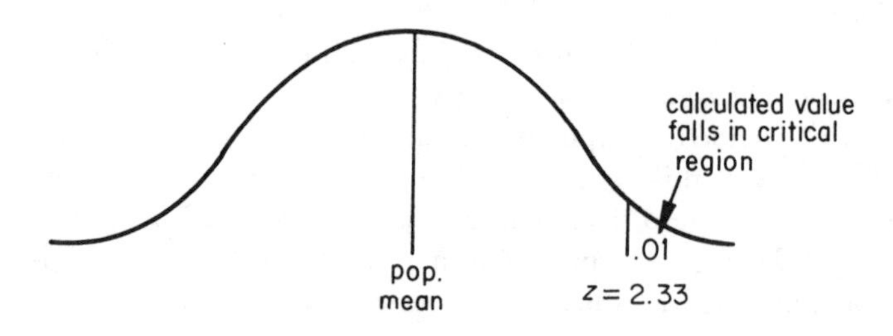

Situation 2: the *t* test—the population standard deviation is unknown.

Let us examine a situation in which we have a randomly selected sample of 20 cases. This is a sample of students who have participated in an enriched learning program. Their mean score on a subject mastery test is 87.62 with a standard deviation of 6.19. The mean score of the general student body is 84.35.

We want to know if the enriched learning program increases subject mastery. Other ways of saying this are either (1) *two-tailed test*: we want to know if there is a significant *difference* in the mean subject mastery of students who participated in the enriched learning program and the mean subject mastery of the general student body (i.e. does the sample come from a population with a different mean?), or (2) *one-tailed test*: we want to know if students who participated in the enriched learning program have a significantly *higher* mean subject mastery score than the general student body.

To set up a test of a hypothesis, we go through the following steps:

1) Construct a *null hypothesis*. The null hypothesis here should state that the sample mean comes from the population's distribution of sample means. In symbolic form, it looks like this: $\mu = \mu_0$.

2) Construct an *alternative* or *research hypothesis*. The alternative or research hypothesis should state either that (a) the sample mean comes from a *different* population. This is a *two-tailed* test. In symbolic form, the hypothesis looks like this: $\mu \neq \mu_0$; or that (b) the sample mean comes from a population with a mean *greater than* (or *less than*) the population mean described here. This is a *one-tailed test*. In symbolic form, it looks like this: $\mu > \mu_0$ for "greater than" (or $\mu < \mu_0$ for "less than"). In this case, let us choose a two-tailed hypothesis: $\mu \neq \mu_0$.

3) Choose a *test statistic*. Select the one-sample t test because you have a sample which has been randomly selected and because we know the sample mean and standard deviation, the population mean and the number of cases in the sample. We do not know the population standard deviation in this example.

4) Choose a *significance level*. We must choose either 0.05, 0.01 or 0.001 for the tail(s). The larger your significance level (0.05 is the largest), the greater your chance of rejecting your null hypothesis. Let us select 0.05 for this example.

5) Select a *sampling distribution*. With t tests, the sampling distribution is always the t distribution. t distributions vary, depending upon the number of cases in the sample. Therefore, you must specify the *degrees of freedom* (df) of the t distribution you are using. The degrees of freedom are obtained by subtracting 1 from the number of cases in the sample. In this case, $df = N - 1 = 20 - 1 = 19$.

6) Select a *critical value*. We select a critical value with which we will compare our *calculated value*, obtained using a t ratio. If our calculated value is equal to or greater than our critical value, our t value falls in a tail of the distribution. We then reject our null hypothesis and accept our research hypothesis. We find our critical value in the t distribution table at the end of this chapter. Look up your critical value according to (a) the degrees of freedom in your distribution; (b) the probability level you have selected and (c) whether or not your test is one-tailed or two-tailed. With 19 degrees of freedom, at the 0.05 level of probability and a two-tailed test, for example, the critical value is 2.093.

7) We now *carry out our test*. The main purpose of the one-

sample *t* test is to calculate if the sample mean comes from a population that is different from the population represented in the problem.

To carry out a one-sample *t* test, do the following:

a) Subtract the population mean from the sample mean. In this case, 87.62 – 84.35 = 3.27.

b) Subtract 1 from the number of cases in the sample. In this case, 20 – 1 = 19.

c) Compute the square root of the number of cases in the sample minus one. The square root of 19 is 4.36.

d) Divide the sample standard deviation by the result of step 7c. In this case, 6.19/4.36 = 1.420.

e) Divide the result of step a by the result of step 7d. This is your calculated *t* value.* In this case, 3.27/1.420 = 2.303.

These steps are based upon the following formula for calculating a *t* value when comparing a sample mean to a population mean:

$$t = \frac{\text{sample mean} - \text{population mean}}{\text{sample standard deviation} / \sqrt{\text{number of cases in sample} - 1}}$$

or, in symbolic form,

$$t = \frac{\overline{X} - \mu}{s / \sqrt{N - 1}} = \frac{87.62 - 84.35}{6.19 / \sqrt{20 - 1}} = 2.303$$

8) We compare 2.303 (our calculated value) to the critical value of 2.093. Since our calculated value is higher than the critical value, the decision we make is to reject our null hypothesis and accept our research hypothesis. We conclude that the mean subject mastery score of students who have participated in an enriched learning program is significantly

*If your calculated value has a negative sign, do not worry. This happens whenever your sample mean is less than your population mean. This is only a problem if you have set up a one-tailed test which predicted that your sample mean would be significantly greater than your population mean and have therefore set up a critical region on the wrong side of the normal curve.

higher than the mean subject mastery score of the general student body.

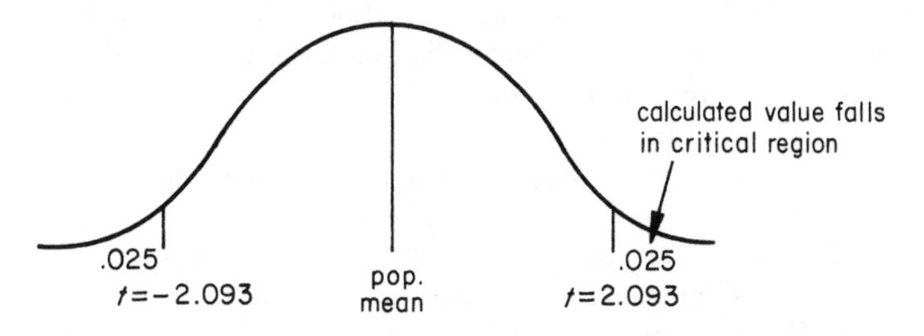

Review Questions

1. What are the two types of hypotheses? What are their differences?
2. What is meant by the tails of a distribution? What is the difference between a one-tailed and a two-tailed test?
3. What is the advantage of a one-tailed test over a two-tailed test?
4. What are the commonly used significance levels? Which one allows you the greatest chance of rejecting your null hypothesis? Which one provides the greatest chance of making a type I error?
5. When is a t test used instead of a z test?
6. What information must you have about your sample and population to do a one-sample z test? To do a one-sample t test?
7. How do you compute the degrees of freedom for a one-sample t test?
8. If your calculated value is less than the critical value, what decision should you make when testing a hypothesis?
9. What is meant by the standard error?

Review Exercises

Exercise 1. A market researcher wants to know if the popularity of blue jeans with college students has changed since last year. In a random sample of 36 college students, the mean number of pairs of blue jeans owned is 3.5.

Test the hypothesis that the population mean is still 3.0 (last year's mean) with a standard deviation of 1.0. What should the market researcher's conclusion be?

1) The null hypothesis: $\mu = \mu_0$
2) The research hypothesis: $\mu \neq \mu_0$ (The test will be two-tailed.)
3) The test statistic: a one-sample z test
4) The significance level: 0.05
5) The sampling distribution: the normal probability curve
6) The critical value: 1.96
7) To carry out our test:
   a) The sample mean minus the population mean $= 3.5 - 3.0 = .5$.
   b) The square root of the number of cases in the sample $=$ the square root of $36 = 6$
   c) The population standard deviation divided by the result of step 7b $= 1.0/6 = .167$.
   d) The result of step 7a divided by the result of step 7c $= .5/.167 = 2.99$.
8) The calculated value of 2.99 is greater than the critical value of 1.96. Therefore, we reject the null hypothesis and accept the research hypothesis. We conclude that the popularity of blue jeans with college students has changed since last year.

Exercise 2. The management of a business which manufactures sports equipment wishes to know if the amount of money consumers spend on recreation has increased since last year. In a random sample of 100, the mean amount spent per week is $50. Test the hypothesis that the population mean is still $45 (last year's mean) with a standard deviation of $10. Draw a conclusion.

1) The null hypothesis: $\mu = \mu_0$
2) The research hypothesis: $\mu > \mu_0$ (The test will be one-tailed.)
3) The test statistic: a one-sample z test
4) The significance level: 0.01

5) The sampling distribution: the normal probability curve
6) The critical value: 2.33
7) To carry out our test:
a) The sample mean minus the population mean = 50 – 45 = 5.
b) The square root of the number of cases in the sample = $\sqrt{100} = 10$.
c) The population standard deviation divided by the result of step 7b = 10/10 = 1.
d) The result of step 7a divided by the result of step 7c = 5/1 = 5.0.
8) The calculated value of 5.0 is greater than the critical value of 2.33. Therefore, we reject the null hypothesis and accept the research hypothesis. We conclude that the amount of money consumers spend on recreation has increased since last year.

Exercise 3. A local crime reporter wishes to test the hypothesis that the rate of violent crime by women has increased in the city in the past year. The average crime rate for women in all of the city's precincts last year was 7 crimes per 100 women.

In a random sample of 20 precincts, the researcher finds an average crime rate by women of 10 crimes per 100 women with a standard deviation of 2.5. Test the reporter's hypothesis and draw a conclusion.

1) The null hypothesis: $\mu = \mu_0$
2) The research hypothesis: $\mu > \mu_0$ (The test will be one-tailed.)
3) The test statistic: a one-sample *t* test
4) The significance level: 0.01
5) The sampling distribution: the *t* distribution with 19 degrees of freedom (df = N – 1 = 20 – 1 = 19)
6) The critical value: 2.539
7) To carry out our test:
a) The sample mean minus the population mean = 10 – 7 = 3.
b) The number of cases in the sample minus 1 = 20 – 1 = 19.
c) The square root of the result of step 7b = $\sqrt{19} = 4.36$.

d) The sample standard deviation divided by the result of step 7c = 2.5/4.36 = .573.

e) The result of step 7a divided by the result of step 7d = 3/.573 = 5.236.

8) The calculated value of 5.236 is greater than the critical value of 2.539. Therefore we reject the null hypothesis and accept the research hypothesis. We conclude that the rate of violent crime by women has increased in the city in the past year.

Exercise 4. In a study of 15 urban communities with street cleaning campaigns, the mean littering index was found to be 34.03 with a standard deviation of 4.80. Test the hypothesis that this littering index is different from the general urban mean littering index of 36.10 and draw a conclusion.

1) The null hypothesis: $\mu = \mu_0$

2) The research hypothesis: $\mu \neq \mu_0$ (The test will be two-tailed.)

3) The test statistic: a one-sample t-test

4) The significance level: 0.05

5) The sampling distribution: the t distribution with 14 degrees of freedom (df = N − 1 = 15 − 1 = 14).

6) The critical value: 2.145

7) To carry out our test:

a) The sample mean minus the population mean = 34.03 − 36.10 = − 2.07.

b) The number of cases in the sample minus 1 = 15 − 1 = 14.

c) The square root of the result of step 7b = $\sqrt{14} = 3.74$.

d) The sample standard deviation divided by the result of step 7c = 4.80/3.74 = 1.283.

e) The result of step 7a divided by the result of step 7d = − 2.07/1.283 = − 1.613.

8) The calculated value of − 1.613 is less than the critical value of 2.145. Therefore, we do not reject our null hypothesis. We conclude that urban communities with street cleaning campaigns do not have a mean littering index different from that of urban communities in general.

Table 8.1 *t* Distribution Table

df	one-tailed test probability values		two-tailed test probability values	
	0·05	0·01	0·05	0·01
1	6.314	31.821	12.706	63.657
2	2.920	6.965	4.303	9.925
3	2.353	4.541	3.182	5.841
4	2.132	3.747	2.776	4.604
5	2.015	3.365	2.571	4.032
6	1.943	3.143	2.447	3.707
7	1.895	2.998	2.365	3.499
8	1.860	2.896	2.306	3.355
9	1.833	2.821	2.262	3.250
10	1.812	2.764	2.228	3.169
11	1.796	2.718	2.201	3.106
12	1.782	2.681	2.179	3.055
13	1.771	2.650	2.160	3.012
14	1.761	2.624	2.145	2.977
15	1.753	2.602	2.131	2.947
16	1.746	2.583	2.120	2.921
17	1.740	2.567	2.110	2.898
18	1.734	2.552	2.101	2.878
19	1.729	2.539	2.093	2.861
20	1.725	2.528	2.086	2.845
21	1.721	2.518	2.080	2.831
22	1.717	2.508	2.074	2.819
23	1.714	2.500	2.069	2.807
24	1.711	2.492	2.064	2.797
25	1.708	2.485	2.060	2.787
26	1.706	2.479	2.056	2.779
27	1.703	2.473	2.052	2.771
28	1.701	2.467	2.048	2.763
29	1.699	2.462	2.045	2.756
30	1.697	2.457	2.042	2.750
40	1.684	2.423	2.021	2.704
60	1.671	2.390	2.000	2.660
120	1.658	2.358	1.980	2.617
∞	1.645	2.326	1.960	2.576

Source: Adapted from Table III of Fisher and Yates, *Statistical Tables for Biological. Agricultural and Medical Research*, published by Longman Group, Ltd., London (previously published by Oliver and Boyd, Ltd., Edinburgh). Used by permission of the authors and publishers.

9

Two-sample z and t Tests

How to Compare the Mean of One Sample to the Mean of a Second Sample

In Chapter 8, you learned how to compare a sample mean to a population mean to see if the sample is likely to have come from that population or a different one. In this chapter, you will learn how to compare a sample mean to a second mean to see if the samples are likely to have come from the same population or different populations. To do this, we need to know (1) the mean of each sample, (2) either the variance or standard deviation of each sample, or of each population,* and (3) the number of cases in each sample. Again, it is assumed that each sample has been taken from a population that is normally distributed.

Comparing the Means of Two Independently Selected Samples

Situation 1: the z test: the population variance of each sample is known or the number of cases in each sample is large (30 or more cases in each).

The population variance of each is rarely known. If you do happen to know the population variances, use them. If not,

*Remember that the variance is the standard deviation squared.

sample variances can be substituted as estimates of population variances *if* both samples are large.

Let us examine a situation in which we have two independently selected samples, each with 100 cases. The first group is a random sample of 100 working mothers with access to day-care centers. The second group is a random sample of 100 working mothers who do not have access to day-care centers. The mean job satisfaction score of the day-care mothers is 15.35 with a variance of 5.29. The mean job satisfaction score of the non-day-care mothers is 14.23 with a variance of 4.78. We want to know if access to day-care centers affects the job satisfaction of working mothers. Other ways of saying this are (1) *two-tailed test*: we want to know if there is a significant *difference* in the mean job satisfaction scores of both groups (i.e., do the two samples come from two populations with different means) or (2) *one-tailed test*: we want to know if the mean job satisfaction score of one group is significantly *higher* than that of the other group.

To set up a test of a hypothesis, we go through the same set of steps we used previously in Chapter 8:

1) Construct a *null hypothesis*. The null hypothesis here should state that the two samples come from populations with the same mean. In symbolic form, it looks like this: $\mu_1 = \mu_2$.

2) Construct an *alternative or research hypothesis*. The alternative or research hypothesis should state either that (a) the two samples come from two populations with *different* means. This is a *two-tailed* test. In symbolic form, the hypothesis looks like this: $\mu_1 \neq \mu_2$. Or (b) the mean of one population is *higher* than the mean of the other population. This is a *one-tailed* test. In symbolic form, the hypothesis looks like this: $\mu_1 > \mu_2$ or $\mu_2 > \mu_1$ depending upon which population, the first or the second, was thought to have the higher mean. For this example, let us choose a one-tailed hypothesis: $\mu_1 > \mu_2$.

3) Choose a *test statistic*. Select the two-sample z test because we have two *large* samples, each of which has been taken from a population that is randomly distributed and because we know the mean, variance and number of cases in each sample, even if we do not know the population variances.

4) Choose a *significance level* (either 0.05, 0.01 or 0.001).

Remember that the larger your significance level (0.05 is the largest), the greater your chance of rejecting your null hypothesis. Let us select 0.05 for this example.

5) Select a *sampling distribution.* With z tests, the sampling distribution is always the normal probability curve.

6) Select a *critical value.* As with our other statistics for hypothesis testing, we select a critical value with which we will compare our *calculated value.* As with other critical values, if our calculated value is equal to or greater than our critical value we reject our null hypothesis and accept our research hypothesis. We find our critical values in the normal z distribution table. As in Chapter 8, the following are the critical z values for various significance levels with both one-tailed and two-tailed hypotheses:

One-tailed Values

significance level	critical z value
0.05	1.65
0.01	2.33
0.001	3.08

Two-tailed Values

significance level	critical z value
0.05	1.96
0.01	2.58
0.001	3.30

For a one-tailed test at the 0.05 level, therefore, our critical z value is 1.65.

7) We now *carry out our test.* The main purpose of the two-sample z test is to calculate if the mean of one independent sample represents a population that is substantially different from the population represented by the mean of the second independent sample.

To carry out a two-sample z test, do the following:

a) Subtract the mean of the second sample from the mean of the first sample. In this case, $15.35 - 14.23 = 1.12$.

b) Divide the variance of the first population (or sample) by the number of cases in the first sample. In this case, 5.29/100 = .0529.

c) Divide the variance of the second population (or sample) by the number of cases in the second sample. In this case 4.78/100 = .0478.

d) Add the result of step 7b to the result of step 7c. In this case, .0529 + .0478 = .1007.

e) Compute the square root of the result of step 7d. In this example, the square root of .1007 is .32.

f) Divide the result of step 7a by the result of step 7e. This is your calculated z value.* In this case, 1.12/.32 = 3.50.

These steps are based upon the following formula for calculating a z value when comparing two sample means:

$$z = \frac{\text{mean of first sample} - \text{mean of second sample}}{\sqrt{\dfrac{\text{variance of first population}}{\text{number of cases in first sample}} + \dfrac{\text{variance of second population}}{\text{number of cases in second sample}}}}$$

or, in symbolic form,

$$z = \frac{\bar{X}_1 - \bar{X}_2}{\sqrt{\dfrac{\sigma_1^2}{N_1} + \dfrac{\sigma_2^2}{N_2}}} = \frac{15.35 - 14.23}{\sqrt{\dfrac{5.29}{100} + \dfrac{4.78}{100}}} = 3.50$$

8) We compare 3.50 (our calculated value) to the critical value of 1.65. Since our calculated value is higher than the critical value the decision we make is to reject our null hypothesis and accept our research hypothesis. We conclude that the mean job satisfaction of working mothers who have access to day-care

*Do not worry if your calculated z has a negative value. That happens whenever the mean of the second sample is higher than the mean of the first sample. It is the absolute value which is important here, unless you have set up a one-tailed test which predicts that your first sample mean is higher than the second sample mean and have therefore set up a critical region on the wrong end of the normal curve.

centers is greater than the mean job satisfaction of working mothers who do not have access to day-care centers.

Situation 2: The t test: the population variance of each sample is unknown.*

Let us examine a situation in which we have two independently selected samples, each with 10 cases. The first group is a random sample of 10 students who are given a drug before having to learn a list of nonsense words. The second group is a random sample of 10 students who are given a placebo before having to learn to list of nonsense words. The mean word retention score of the drug group is 9.06 with a standard deviation of 3.22.

The mean word retention score of the placebo group is 12.83 with a standard deviation of 4.94. We want to know if exposure to the drug affects performance on the word memorization task. Other ways of saying this are (1) *two-tailed test*: we want to know if there is a significant *difference* in the mean performance scores of both groups, or (2) *one-tailed test*: we want to know if the mean performance score of one group is significantly *higher* than the mean performance score of the other group.

To set up a test of a hypothesis, we go through the same set of steps we used previously in Chapter 8:

1) Construct a *null hypothesis*. The null hypothesis here should state that the two samples come from populations with the same mean. In symbolic form, it looks like this: $\mu_1 = \mu_2$.

2) Construct an *alternative* or *research hypothesis*. The alternative or research hypothesis should state either that (a) the two samples come from populations with *different* means. This is a *two-tailed* test. In symbolic form, the hypothesis looks like this: $\mu_1 \neq \mu_2$. Or (b) the mean of one population is higher than the mean of the other population. This is a *one-tailed* test. In symbolic form, the hypothesis looks like this: $\mu_1 > \mu_2$ or $\mu_2 > \mu_1$. For this example, let us choose a two-tailed hypothesis: $\mu_1 \neq \mu_2$.

3) Choose a *test statistic*. Select the two-sample t test because

*Although this method will not be exact if the numbers of cases in both samples are very dissimilar, in most small sample situations, the two samples being compared are reasonably close in size. Remember that the variance is the standard deviation squared.

we have small samples which have been randomly selected and because we know the mean, variance, and number of cases in each sample, and we *do not know* the population variances.

4) Choose a *significance level* (either 0.05, 0.01 or 0.001). Remember that the larger your significance level (0.05 is the largest), the greater your chance of rejecting your null hypothesis. Let us select 0.01 for this example.

5) Select a *sampling distribution*. With t tests, the sampling distribution is always the t distribution. t distributions vary, depending upon the number of cases in each sample. Therefore, you must specify the *degrees of freedom* (*df*) of the t distribution you are using. The degrees of freedom are obtained by finding $N_1 + N_2 - 2$. In other words, add the number of cases in both samples together and subtract 2. In this case, $df = 10 + 10 - 2 = 20 - 2 = 18$.

6) Select a *critical value*. As with our other statistics for hypothesis testing, we select a critical value with which we will compare our *calculated* value. If our calculated value is equal to or greater than our critical value, we reject our null hypothesis and accept our research hypothesis. We find our critical value in the t distribution table at the end of this chapter. This is the same table as you find at the end of Chapter 8. Look up your critical value according to (a) the degrees of freedom in your distribution, (b) the probability level you have selected, and (c) whether your test is one-tailed or two-tailed. With 18 degrees of freedom, at the 0.01 level of probability and a two-tailed test, for example, the critical value is 2.878.

7) We now *carry out our test*. The main purpose of the two-sample t test is to calculate if the mean of one independent sample represents a population that is substantially different from the population represented by the second independent sample.

To carry out a two-sample t test, do the following:

a) Subtract the mean of the second sample from the mean of the first sample. In this case, $9.06 - 12.83 = -3.77$.

b) Subtract 1 from the number of cases in the first sample. In this case, $10 - 1 = 9$.

c) Find the square root of the result of step 7b. In this case, the square root of 9 is 3.

d) Divide the standard deviation of the first sample by the result of step 7c. In this example, 3.22/3 = 1.073.

e) Square the result of step 7d. In this case, $(1.073)^2 =$ 1.151.

f) Subtract 1 from the number of cases in the second sample. In this case, 10 – 1 = 9.

g) Find the square root of the result of step 7f. In this case, the square root of 9 is 3.

h) Divide the standard deviation of the second sample by the result of step 7g. In this case, 4.94/3 = 1.647.

i) Square the result of step 7h. In this case, $(1.647)^2 =$ 2.713.

j) Add the result of steps 7e and 7i. In this case 1.151 + 2.713 = 3.864.

k) Find the square root of the result of step 7j. In this case, the square root of 3.864 is 1.97.

l) Divide the result of step 7a by the result of step 7k. This is your calculated t value. Do not worry about the negative value. That happens whenever the mean of the second sample is higher than the mean of the first sample. It is the absolute value which is important here, unless you have predicted in a one-tailed test that the mean of the second sample is lower than the mean of the first sample. In this case, – 3.77/1.97 = – 1.914.

These steps are based upon the following formula for calculating a t value when comparing two sample means:

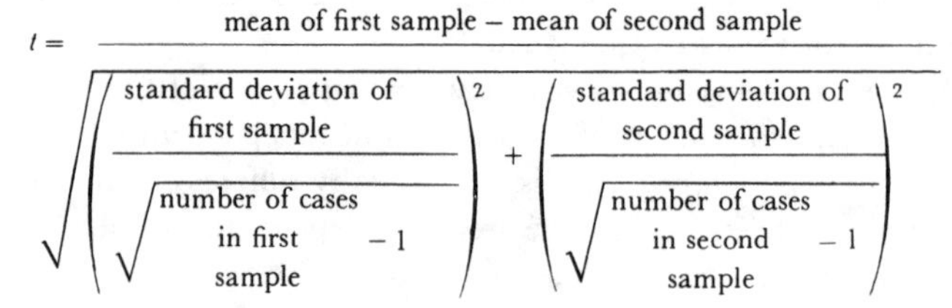

$$t = \frac{\text{mean of first sample} - \text{mean of second sample}}{\sqrt{\left(\frac{\text{standard deviation of first sample}}{\sqrt{\text{number of cases in first sample} - 1}}\right)^2 + \left(\frac{\text{standard deviation of second sample}}{\sqrt{\text{number of cases in second sample} - 1}}\right)^2}}$$

or, in symbolic form,

$$t = \frac{\bar{X}_1 - \bar{X}_2}{\sqrt{\left(\frac{s_1}{\sqrt{N_1 - 1}}\right)^2 + \left(\frac{s_2}{\sqrt{N_2 - 1}}\right)^2}} = \frac{9.06 - 12.83}{\sqrt{\left(\frac{3.22}{\sqrt{10 - 1}}\right)^2 + \left(\frac{4.94}{\sqrt{10 - 1}}\right)^2}} = -1.914$$

8) We compare 1.914 (our calculated value) to the critical value (the distribution table value) of 2.878. Since our calculated value is less than the critical value, the decision we make is not to reject our null hypothesis. We conclude that there is no difference between the mean performance scores of the drug and the placebo group.

Comparing Two Dependent Samples

Sometimes two samples consist of pairs of individuals who have been *matched* on selected characteristics (they are similar on those characteristics). Samples may also be dependent if the same cases are measured before a stimulus (sample 1) and after a stimulus (sample 2).

If we want to examine differences between two *dependent* samples, we must treat each *pair* of cases as a single case.

Suppose that we have ten pairs of individuals matched for income, residence type and family size. One set of ten individuals is exposed to a film favoring scattered site public housing. The other set of ten views a neutral film on this subject. We wish to know if there is a difference between the two groups' attitudes toward scattered site public housing after the films. The following are the scores of both groups on an attitude scale measuring the extent to which scattered site public housing is favored:

Pair	Sample 1 Pro housing film	Sample 2 neutral film
A	81	77
B	83	86
C	49	51
D	72	69
E	94	90
F	35	32
G	21	18
H	15	11
I	52	46
J	12	14

We set up our hypothesis testing procedure in the same manner as a two-sample t test on independently selected samples and make the same assumtions.

The differences are: 1) The degrees of freedom used in the selection of a sampling distribution equal the *number of pairs-1.* In this case, df = 10 – 1 = 9.

2) The test statistic used differs from the two-sample t test. It is called a *matched-pairs t* test.

To carry out our test, do the following:

a) Make a fourth column labeled "D" or "difference." Compute the difference between the scores for each matched pair and place the results in the "D" column. In this case, your columns should now look like this:

Pair	Sample 1 Pro housing film	Sample 2 neutral film	D
A	81	77	4
B	83	86	– 3
C	49	51	– 2
D	72	69	3
E	94	90	4
F	35	32	3
G	21	18	3
H	15	11	4
I	52	46	6
J	12	14	– 2

b) Make a fifth column labeled D^2. Square each number in the D column and enter it in the D^2 column. Your columns should now look like this:

Pair	Sample 1 Pro housing film	Sample 2 neutral film	D	D^2
A	81	77	4	16
B	83	86	– 3	9
C	49	51	– 2	4
D	72	69	3	9

E	94	90	4	16
F	35	32	3	9
G	21	18	3	9
H	15	11	4	16
I	52	46	6	36
J	12	14	– 2	4
			$\sum = 20$	$\sum = 128$

c) Add up the numbers in the "D" column. (Notice that some numbers may be negative.) In this case, the numbers sum to 20.

d) Divide the result of step 7c by the number of pairs of scores. This is your mean of sample differences. In this example, $20/10 = 2$.

e) Add up the numbers in the "D^2" column. In this case the sum of $D^2 = 128$.

f) Divide the result of step 7e by the number of pairs of scores. In this example, $128/10 = 12.8$.

g) Square the result of step 7d. In this case, 2 squared is 4.

h) Subtract the result of step 7g from the result of step 7f. In this case, $12.8 - 4 = 8.8$.

i) Compute the square root of the result of step 7h. This is the standard deviation of differences. In this case, the square root of 8.8 is 2.97.

j) Subtract 1 from the number of pairs of scores. In this case, $10 - 1 = 9$.

k) Compute the square root of the result of step 7j. In this case, the square root of 9 is 3.

l) Divide the result of step 7i by the result of step 7k. In this case $2.97/3 = .99$.

m) Divide the result of step 7d by the result of step 7l. This is your calculated *t* value. In this case, $2/.99 = 2.02$. Whether or not the null hypothesis would be rejected would depend upon the significance level chosen and whether the test were one-tailed or two-tailed.

These steps for calculating a matched-pairs *t* value are based upon the following formula:

$$t = \frac{\text{mean of the differences}}{\dfrac{\text{standard deviation of the differences}}{\sqrt{\text{number of pairs of cases} - 1}}}$$

or, in symbolic form,

$$t = \frac{\overline{X}_D}{S_D \big/ \sqrt{N-1}} = \frac{2}{2.97 \big/ \sqrt{10-1}} = 2.02$$

Review Questions

1. When is a t test used instead of a z test?
2. When is a matched-pairs t test used?
3. What assumptions are made about the samples and the population (s) in two-sample z and t tests?
4. What information must you have about your samples to do a two-sample z test? To do a two-sample t test?
5. How do you compute the degrees of freedom for a two-sample t test?
6. What is the difference between a sample and a population?

Review Exercises

Exercise 1. A researcher is interested in wage differences of unionized workers in two different industries. In the first industry, a random sample of 49 workers revealed a mean hourly wage of $14.77, with a standard deviation of $2.96. In the second industry, a random sample of 49 workers revealed a mean hourly wage of $13.72, with a standard deviation of $2.57. Test the hypothesis that mean hourly wages of unionized workers are different in the two industries and draw a conclusion.

1) The null hypothesis: $\mu_1 = \mu_2$

2) The research hypothesis: $\mu_1 \neq \mu_2$ (The test will be two-tailed.)

3) The test statistic: a two-sample z test

4) The significance level: 0.01

5) The sampling distribution: the normal probability curve

6) The critical value: 2.58

7) To carry out our test:

a) The mean of the first sample minus the mean of the second sample = 14.77 – 13.72 = 1.05.

b) The variance (i.e., standard deviation squared) of the first sample divided by the number of cases in the first sample = $(2.96)^2/49 = 8.76/49 = .18$.

c) The variance (i.e., standard deviation squared) of the second sample divided by the number of cases in the second sample = $(2.57)^2/49 = 6.60/49 = .13$

d) The sum of the result of step 7b and the result of step 7c = .18 + .13 = .31.

e) The square root of the result of step 7d = $\sqrt{.31} = .56$.

f) The result of step 7a divided by the result of step 7e = 1.05/.56 = 1.88.

8) The calculated value of 1.88 is less than the critical value of 2.58. Therefore we do not reject the null hypothesis. We conclude that mean hourly wages of unionized workers are not different in the two industries.

Exercise 2. Sixteen people were selected and randomly assigned to either high anxiety or low anxiety work conditions (eight people in each condition) in order to measure the effects of anxiety on sociability in the office. The high anxiety group had a mean sociability rating of 1.88 with a standard deviation of .78. The low anxiety group had a mean sociability rating of 2.63 with a standard deviation of .86. Test the hypothesis that high anxiety workers are less sociable and draw a conclusion.

1) The null hypothesis: $\mu_1 = \mu_2$

2) The research hypothesis: $\mu_2 > \mu_1$ (The test will be one-tailed.)

3) The test statistic: a two-sample t test.

4) The significance level: 0.05

5) The sampling distribution: the t distribution with 14 degrees of freedom ($df = N_1 + N_2 - 2 = 8 + 8 - 2 = 14$)

6) The critical value: 1.761

7) To carry out our test:

a) The mean of the first sample minus the mean of the second sample $= 1.88 - 2.63 = -.75$.

b) The number of cases in the first sample minus $1 = 8 - 1 = 7$.

c) The square root of the result of step 7b $= \sqrt{7} = 2.65$

d) The standard deviation of the first sample divided by the result of step 7c $= .78/2.65 = .294$.

e) The result of step 7d squared $= (.294)^2 = .086$.

f) The number of cases in the second sample minus $1 = 8 - 1 = 7$.

g) The square root of the result of step 7f $= \sqrt{7} = 2.65$.

h) The standard deviation of the second sample divided by the result of step 7g $= .86/2.65 = .325$.

i) The result of step 7h squared $= (.325)^2 = .106$.

j) The sum of the results of steps 7e and 7i $= .086 + .106 = .192$.

k) The square root of the result of step 7j $= \sqrt{.192} = .44$.

l) The result of step 7a divided by the result of step 7k $= -.75/.44 = -1.705$.

8) The calculated value of -1.705 is less than the critical value of 1.761. Therefore we do not reject our null hypothesis. We conclude that high anxiety workers are not less sociable than low anxiety workers in the office.

Exercise 3. A researcher is interested in finding out whether taller men score higher in dominance than shorter men. A sample of 10 short males and a sample of 10 tall males are selected and matched for age, race, income and education. The following are the dominance ratings:

Pair	Short males	Tall males
A	58	68
B	51	70
C	73	60
D	68	74
E	77	76
F	72	71

G	49	56
H	59	79
I	77	100
J	72	74

Test the researcher's hypothesis and draw a conclusion.

1) The null hypothesis: $\mu_1 = \mu_2$

2) The research hypothesis: $\mu_2 > \mu_1$ (The test will be one-tailed.)

3) The test statistic: a matched-pairs t test.

4) The significance level: 0.05

5) The sampling distribution: the t distribution with 9 degrees of freedom (df = number of pairs minus one = 10 − 1 = 9)

6) The critical value: 1.833

7) To carry out our test:

a) We make a fourth column labeled "D" and compute the difference between the scores for each matched pair:

Pair	Short males	Tall males	D
A	58	68	− 10
B	51	70	− 19
C	73	60	13
D	68	74	− 6
E	77	76	1
F	72	71	1
G	49	56	− 7
H	59	79	− 20
I	77	100	− 23
J	72	74	− 2

b) We make a fifth column labeled "D^2" and square each number from the D column:

Pair	Short males	Tall males	D	D^2
A	58	68	− 10	100
B	51	70	− 19	361
C	73	60	13	169
D	68	74	− 6	36

E	77	76	1	1
F	72	71	1	1
G	49	56	− 7	49
H	59	79	− 20	400
I	77	100	− 23	529
J	72	74	− 2	4

c) The sum of the numbers in the D column = − 72.

d) The result of step 7c divided by the number of pairs of scores = − 72/10 = − 7.2.

e) The sum of the numbers in the D^2 column = 1650.

f) The result of step 7e divided by the number of pairs of scores = 1650/10 = 165.

g) The result of step 7d squared = $(-7.2)^2 = 51.84$.

h) The result of step 7f minus the result of step 7g = 165 − 51.84 = 113.16.

i) The square root of the result of step 7h = $\sqrt{113.16} = 10.64$.

j) The number of pairs of scores minus 1 = 10 − 1 = 9.

k) The square root of the result of step 7j = $\sqrt{9} = 3$.

l) The result of step 7i divided by the result of step 7k = 10.64/3 = 3.547.

m) The result of step 7d divided by the result of step 7l = − 7.2/3.547 = − 2.030.

8) The calculated value of − 2.030 is higher than the critical value of 1.833 (It is the absolute value which is important here, not the signs.). Therefore we reject the null hypothesis and accept the research hypothesis. We conclude that taller men score higher in dominance than shorter men.

Table 9.1 t Distribution Table

	one-tailed test probability values		two-tailed test probability values	
	0.05	0.01	0.05	0.01
1	6.314	31.821	12.706	63.657
2	2.920	6.965	4.303	9.925
3	2.353	4.541	3.182	5.841
4	2.132	3.747	2.776	4.604
5	2.015	3.365	2.571	4.032
6	1.943	3.143	2.447	3.707
7	1.895	2.998	2.365	3.499
8	1.860	2.896	3.306	3.355
9	1.833	2.821	2.262	3.250
10	1.812	2.764	2.228	3.169
11	1.796	2.718	2.201	3.106
12	1.782	2.681	2.179	3.055
13	1.771	2.650	2.160	3.012
14	1.761	2.624	2.145	2.977
15	1.753	2.602	2.131	2.947
16	1.746	2.583	2.120	2.921
17	1.740	2.567	2.110	2.898
18	1.734	2.552	2.101	2.878
19	1.729	2.539	2.093	2.861
20	1.725	2.528	2.086	2.845
21	1.721	2.518	2.080	2.831
22	1.717	2.508	2.074	2.819
23	1.714	2.500	2.069	2.807
24	1.711	2.492	2.064	2.797
25	1.708	2.485	2.060	2.787
26	1.706	2.479	2.056	2.779
27	1.703	2.473	2.052	2.771
28	1.701	2.467	2.048	2.763
29	1.699	2.462	2.045	2.756
30	1.697	2.457	2.042	2.750
40	1.684	2.423	2.021	2.704
60	1.671	2.390	2.000	2.660
120	1.658	2.358	1.980	2.617
∞	1.645	2.326	1.960	2.576

Source: Adapted from Table III of Fisher and Yates, *Statistical Tables for Biological, Agricultural and Medical Research*, published by Longman Group, Ltd., London (previously published by Oliver and Boyd, Ltd., Edinburgh). Used by permission of the authors and publishers.

10

Simple Analysis of Variance

How to Compare More Than Two Samples

In Chapter 9, we saw how to compare *two* samples to see if they were likely to have come either from the same population or from different populations. Often, we want to compare *more than two* samples. That is when we use a technique called analysis of variance. Analysis of variance compares the variation between each population represented by the samples (*between-group variance*) to the amount of variation within each population represented by the samples (*within-group variance*). It combines within one procedure all combinations of t tests which would otherwise have to be performed and reduces the chances of rejecting the null hypothesis when it is really true. To do a simple analysis of variance, we need to know (1) the mean of each sample,* (2) the variance of each sample, and (3) the number of cases in each sample. Again, it is assumed that each sample has been taken from a population that is normally distributed, and that the populations have the same variance.

*If we have ordinal level data rather than interval or ratio level data, we would do a Kruskal–Wallis one-way analysis of variance which requires that you know sums of ranks and number of cases instead of the mean of each sample. Assumptions of normality or equality of variance are not necessary.

This chapter will illustrate a one-way analysis of variance in which several samples are compared on one variable. (A more complicated form of analysis of variance is two-way analysis of variance which enables you to look at more than one variable at a time and the joint effects of two or more variables.)

Illustrative Problem

Suppose that we have 3 groups of 10 randomly selected from 3 different populations: urban, suburban and rural dwellers. We measure fear of crime on a scale which varies from 10 to 30. The higher the number, the greater the fear of crime. The following are results from the study:

	Group 1 Rural	Group 2 Suburban	Group 3 Urban
Sample Mean	17.3	22.4	25.1
Sample Variance	3.5	5.1	6.8
Number of cases in sample	10	10	10

We want to know if there are differences in mean fear scores among the three populations.

To set up a test of a hypothesis, we go through the same set of steps we used previously in Chapters 8 and 9:

1) Construct a *null hypothesis*. The null hypothesis here should state that the samples come from populations with the same mean. In symbolic form, it looks like this: $\mu_1 = \mu_2 = \ldots = \mu_k$

2) Construct an *alternative* or *research hypothesis*. The alternative or research hypothesis should state that at least one of the samples comes from a population with a mean higher than or lower than the others. All analysis of variance tests take place in *only the right side tail* of the distribution and are therefore *one-tailed*.

3) Choose a *test statistics*. Select the one-way analysis of variance because we are comparing one variable in more than

two samples. Each sample has been taken from a population that is normally distributed and the populations have the same variance. (So we assume. We know only the sample variances, which are different.)

4) Choose a *significance level* (either 0.05, 0.01 or 0.001). Remember that the larger your significance level (0.05 is the largest) the greater your chance of rejecting your null hypothesis. Let us select 0.05 for this example.

5) Select a *sampling distribution*. With analysis of variance, the sampling distribution is the F distribution. F distributions vary, depending upon the number of samples and the total number of cases in the samples. Therefore, you must specify two types of *degrees of freedom* for the F distribution you are using:

a) The first degrees of freedom (df_1) is obtained by subtracting 1 from the number of samples. In symbolic form, $df_1 = k - 1$. k means the number of samples. In this case, $df_1 = 3 - 1 = 2$.

b) The second degrees of freedom (df_2) is obtained by adding together the number of cases in each sample and subtracting the number of samples. In symbolic form, $df_2 = N_1 + N_2 + \ldots + N_k - k$. In this case, $df_2 = 10 + 10 + 10 - 3 = 27$.

6) Select a *critical value*. As with our other statistics for hypothesis testing, we select a critical value with which we will compare our calculated value. If our calculated value is equal to or greater than our critical value, we reject our null hypothesis and accept our research hypothesis. We find our critical value in the F distribution table at the end of this chapter. Look up your critical value according to (a) the first degrees of freedom, df_1, at the top of the table, (b) the second degrees of freedom, df_2, at the side of the table, and (c) the probability level you have selected. With $df_1 = 2$ and $df_2 = 27$, at the .05 level of probability, for example, the critical value is 3.35.

7) We now *carry our our test*. The main purpose of the one-way analysis of variance is to calculate if the ratio of between-group variance to within-group variance is substantially greater than 1.

To carry our a one-way analysis of variance, do the following:

a) Subtract 1 from the number of cases in the first sample. In this case, $10 - 1 = 9$.

b) Multiply the result obtained in step 7a by the variance for the first sample. In this case $(9) \times (3.5) = 31.5$.

c) Repeat steps 7a and 7b for each of the other samples. In this case, for the second sample, $(9) \times (5.1) = 45.9$. For the third sample, $(9) \times (6.8) = 61.2$.

d) Add the results of steps 7b and 7c. In this case, $31.5 + 45.9 + 61.2 = 138.6$.

e) Add the number of cases in each sample. In this case, $10 + 10 + 10 = 30$.

f) Subtract the number of samples from the result of step 7e. In this case, $30 - 3 = 27$.

g) Divide the result of step 7d by the result of step 7f. This is your within-group variance. In this case, $138.6/27 = 5.13$.

h) Multiply the number of cases in the first sample by the mean of the first sample. In this case, $(10) \times (17.3) = 173$.

i) Repeat step 7h for each of the other samples. In this case, for the second sample, $(10) \times (22.4) = 224$. For the third sample, $(10) \times (25.1) = 251$.

j) Add the results of steps 7h and 7i. In this case, $173 + 224 + 251 = 648$.

k) Square the result of step 7j. In this case, $(648)^2 = 419{,}904$.

l) Divide the result of step 7k by the total number of cases in all samples. In this case, $419{,}904/30 = 13{,}996.8$.

m) Square the mean of the first sample. In this case, $(17.3)^2 = 299.29$.

n) Multiply the result obtained in step 7m by the number of cases in the first sample. In this example, $(299.29) \times (10) = 2992.9$.

o) Repeat steps 7m and 7n for each of the other samples. In this case, for the second sample, $(22.4)^2 = 501.76$; $(501.76) \times (10) = 5017.6$. For the third sample, $(25.1)^2 = 630.01$; $(630.01) \times (10) = 6300.1$.

p) Add the results of steps 7n and 7o. In this case, $2992.9 + 5017.6 + 6300.1 = 14{,}310.6$.

q) Subtract the result of step 7l from the result of step 7p. In this case, $14{,}310.6 - 13{,}996.8 = 313.8$.

r) Subtract 1 from the number of samples. In this case, $3 - 1 = 2$.

s) Divide the result of step 7q by the result of step 7r. This is your between-group variance. In this case, $313.8/2 = 156.9$.

t) Divide the result of step 7s by the result of step 7g. This is your calculated F value. In this case, $156.9/5.13 = 30.58$.

These steps are based upon the following formulas for calculating an F value:

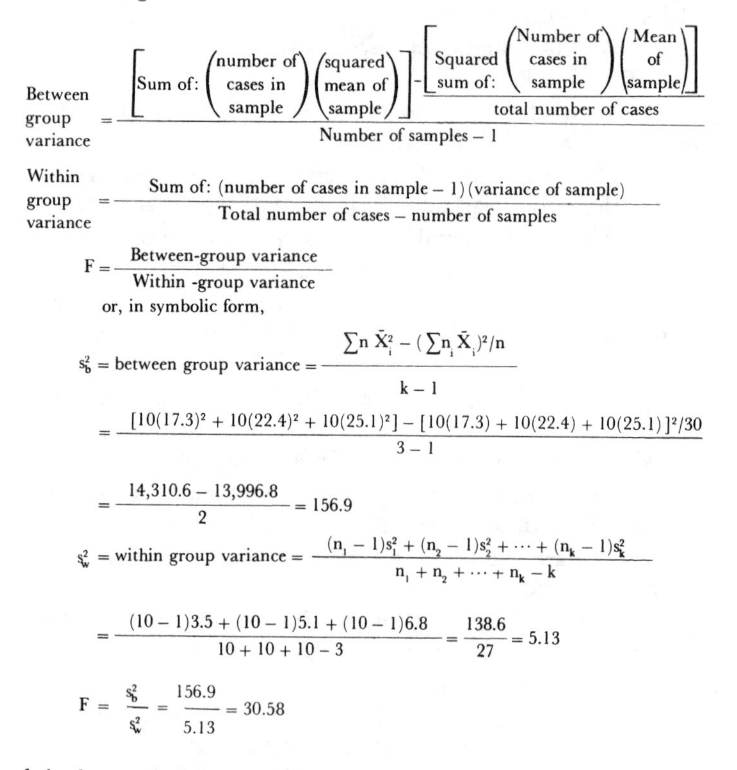

$$\text{Between group variance} = \frac{\left[\text{Sum of: } \begin{pmatrix}\text{number of}\\ \text{cases in}\\ \text{sample}\end{pmatrix}\begin{pmatrix}\text{squared}\\ \text{mean of}\\ \text{sample}\end{pmatrix}\right] - \dfrac{\left[\text{Squared sum of: } \begin{pmatrix}\text{Number of}\\ \text{cases in}\\ \text{sample}\end{pmatrix}\begin{pmatrix}\text{Mean}\\ \text{of}\\ \text{sample}\end{pmatrix}\right]}{\text{total number of cases}}}{\text{Number of samples} - 1}$$

$$\text{Within group variance} = \frac{\text{Sum of: (number of cases in sample} - 1)\text{(variance of sample)}}{\text{Total number of cases} - \text{number of samples}}$$

$$F = \frac{\text{Between-group variance}}{\text{Within -group variance}}$$

or, in symbolic form,

$$s_b^2 = \text{between group variance} = \frac{\sum n\,\bar{X}_i^2 - (\sum n_i \bar{X}_i)^2/n}{k - 1}$$

$$= \frac{[10(17.3)^2 + 10(22.4)^2 + 10(25.1)^2] - [10(17.3) + 10(22.4) + 10(25.1)]^2/30}{3 - 1}$$

$$= \frac{14{,}310.6 - 13{,}996.8}{2} = 156.9$$

$$s_w^2 = \text{within group variance} = \frac{(n_1 - 1)s_1^2 + (n_2 - 1)s_2^2 + \cdots + (n_k - 1)s_k^2}{n_1 + n_2 + \cdots + n_k - k}$$

$$= \frac{(10 - 1)3.5 + (10 - 1)5.1 + (10 - 1)6.8}{10 + 10 + 10 - 3} = \frac{138.6}{27} = 5.13$$

$$F = \frac{s_b^2}{s_w^2} = \frac{156.9}{5.13} = 30.58$$

k is the symbol for number of samples.

8) We compare 30.58 (our calculated value) to the critical value (the distribution table value) of 3.35. Since our calculated value is greater than the critical value, the decision we make is to reject our null hypothesis and accept our research hypothesis. We conclude that the population means are not equal. If we look

at our table of sample data, we can see that rural dwellers have the least fear of crime on the average and urban dwellers have the most fear of crime on the average.

Review Questions

1. What is the advantage of analysis of variance over a series of t tests?
2. What does analysis of variance compare?
3. What information do we need to do a simple analysis of variance?
4. What assumptions are made when using analysis of variance?
5. What sampling distribution do we use with analysis of variance?
6. How do we obtain the first and second types of degrees of freedom of our sampling distribution?

Review Exercises

Exercise 1. A researcher is interested in differences between mean breakdown rates of three different types of additives. Random samples of 8 are selected for each additive type. The following are results from the study:

	Additives		
	A	B	C
Sample mean (in seconds)	24	27	33
Sample Variance	2.7	3.2	4.1
Number of cases in sample	8	8	8

Are there differences in mean breakdown rates among the three additives?

1) The null hypothesis: $\mu_1 = \mu_2 = \mu_3$

2) The research hypothesis: At least one of the samples comes from a population with a mean different from the others.

3) The test statistic: a one-way analysis of variance.

4) The significance level: 0.05.

5) The sampling distribution: the F distribution with $df_1 = 2$ and $df_2 = 21$ (df_1 = number of samples $- 1 = 3 - 1 = 2$; df_2 = total number of cases $-$ number of samples $= 24 - 3 = 21$)

6) The critical value: 3.47

7) To carry out our test:

a) The number of cases in the first sample minus $1 = 8 - 1 = 7$.

b) The result of step 7a multiplied by the variance for the first sample $= (7) \times (2.7) = 18.9$.

c) Steps 7a and 7b repeated for each of the other two samples $= 8 - 1 = 7$; $(7) \times (3.2) = 22.4$ for sample 2. $8 - 1 = 7$; $(7) \times (4.1) = 28.7$ for sample 3.

d) The sum of the results of steps 7b and 7c $= 18.9 + 22.4 + 28.7 = 70.0$.

e) The sum of the number of cases in each sample $= 8 + 8 + 8 = 24$.

f) The result of step 7e minus the number of samples $= 24 - 3 = 21$.

g) The result of step 7d divided by the result of step 7f $= 70.0/21 = 3.33$ (This is the within-group variance).

h) The number of cases in the first sample multiplied by the mean of the first sample $= (8) \times (24) = 192$.

i) Step 7h repeated for each of the other two samples $= (8) \times (27) = 216$ for sample 2. $(8) \times (33) = 264$ for sample 3.

j) The sum of the results of steps 7h and 7i $= 192 + 216 + 264 = 672$.

k) The result of step 7j squared $= (672)^2 = 451{,}584$.

l) The result of step 7k divided by the total number of cases in all samples $= 451{,}584/24 = 18{,}816$.

m) The mean of the first sample squared $= (24)^2 = 576$.

n) The result of step 7m multiplied by the number of cases in the first sample $= (576) \times (8) = 4608$.

o) Steps 7m and 7n repeated for each of the other two samples =

$(27)^2 = 729$; $(729) \times (8) = 5832$ for sample 2. $(33)^2 = 1089$; $(1089) \times (8) = 8712$ for sample 3.

p) The sum of the results of steps 7n and 7o $= 4608 + 5832 + 8712 = 19152$.

q) the result of step 7p minus the result of step 7l $= 19,152 - 18,816 = 336$.

r) The number of samples minus $1 = 3 - 1 = 2$.

s) The result of step 7q divided by the result of step 7r $= 336/2 = 168$ (This is your between-group variance).

t) The result of step 7s divided by the result of step 7g $= 168/3.33 = 50.45$.

8) The calculated value of 50.45 is higher than the critical value of 3.47. Therefore we reject the null hypothesis and accept the research hypothesis. We conclude that at least one of the additives has a breakdown rate different from the others.

Exercise 2. A researcher is interested in audience response to three different kinds of commercials. Random samples of 10 are selected to view each commercial. Each audience member is then given a rating sheet and asked to rate the commercial. The following are results from the study:

	Commercials		
	A	B	C
Sample mean (rating)	8.5	7.3	7.1
Sample Variance	1.2	1.7	2.3
Number of cases in sample	10	10	10

Are there differences in mean ratings for the three commercials?

1) The null hypothesis: $\mu_1 = \mu_2 = \mu_3$

2) The research hypothesis: At least one of the samples comes from a population with a mean different from the others.

3) The test statistic: a one-way analysis of variance

4) The significance level: 0.01

5) The sampling distribution: the F distribution with $df_1 = 2$

and $df_2 = 27$ (df_1 = number of samples − 1 = 3 − 1 = 2; df_2 = total number of cases − number of samples = 30 − 3 = 27)

6) The critical value: 5.49

7) To carry out our test:

a) The number of cases in the first sample minus 1 = 10 − 1 = 9.

b) The result of step 7a multiplied by the variance for the first sample = (9) × (1.2) = 10.8.

c) Steps 7a and 7b repeated for each of the other two samples = 10 − 1 = 9; (9) × (1.7) = 15.3 for sample 2. 10 − 1 = 9; (9) × (2.3) = 20.7 for sample 3.

d) The sum of the results of steps 7b and 7c = 10.8 + 15.3 + 20.7 = 46.8.

e) The sum of the number of cases in each sample = 10 + 10 + 10 = 30.

f) The result of step 7e minus the number of samples = 30 − 3 = 27.

g) The result of step 7d divided by the result of step 7f = 46.8/27 = 1.73 (This is the within-group variance).

h) The number of cases in the first sample multiplied by the mean of the first sample = (10) × (8.5) = 85.

i) Step 7h repeated for each of the other two samples = (10) × (7.3) = 73 for sample 2. (10) × (7.1) = 71 for sample 3.

j) The sum of the results of steps 7h and 7i = 85 + 73 + 71 = 229.

k) The result of step 7j squared = $(229)^2$ = 52,441.

l) The result of step 7k divided by the total number of cases in all samples = 52,441/30 = 1748.03.

m) The mean of the first sample squared $(8.5)^2$ = 72.25.

n) The result of step 7m multiplied by the number of cases in the first sample = (72.25) × (10) = 722.5.

o) Steps 7m and 7n repeated for each of the other two samples = $(7.3)^2$ = 53.29; (53.29) × (10) = 532.9 for sample 2. $(7.1)^2$ = 50.41; (50.41) × (10) = 504.1 for sample 3.

p) The sum of the results of steps 7n and 7o = 722.5 + 532.9 + 504.1 = 1759.5.

q) The result of step 7p minus the result of step 7l = 1759.5 – 1748.03 = 11.47.

r) The number of samples minus 1 = 3 – 1 = 2.

s) The result of step 7q divided by the result of step 7r = 11.47/2 = 5.74.

t) The result of step 7s divided by the result of step 7g = 5.74/1.73 = 3.32.

8) The calculated value of 3.32 is lower than the critical value of 5.49. Therefore we do not reject the null hypothesis. We conclude that audience ratings of the three commercials do not differ.

Table 10.1 F Distribution Table

	df₁											
	p = .05				p = .01				p = .001			
df_2	1	2	3	4	1	2	3	4	1	2	3	4
1	161.4	199.5	215.7	224.6	4052	4999	5403	5625	405284	500000	540379	562500
2	18.51	19.00	19.16	19.25	98.49	99.01	99.17	99.25	998.5	999.0	999.2	999.2
3	10.13	9.55	9.28	9.12	34.12	30.81	29.46	28.71	167.5	148.5	141.1	137.1
4	7.71	6.94	6.59	6.39	21.20	18.00	16.69	15.98	74.14	61.25	56.18	53.44
5	6.61	5.79	5.41	5.19	16.26	13.27	12.06	11.39	47.04	36.61	33.20	31.09
6	5.99	5.14	4.76	4.53	13.74	10.92	9.78	9.15	35.51	27.00	23.70	21.90
7	5.59	4.74	4.35	4.12	12.25	9.55	8.45	7.85	29.22	21.69	18.77	17.19
8	5.32	4.46	4.07	3.84	11.26	8.65	7.59	7.01	25.42	18.49	15.83	14.39
9	5.12	4.26	3.86	3.63	10.56	8.02	6.99	6.42	22.86	16.39	13.90	12.56
10	4.96	4.10	3.71	3.48	10.04	7.56	6.55	5.99	21.04	14.91	12.55	11.28
11	4.84	3.98	3.59	3.36	9.65	7.20	6.22	5.67	19.69	13.81	11.56	10.35
12	4.75	3.88	3.49	3.26	9.33	6.93	5.95	5.41	18.64	12.97	10.80	9.63
13	4.67	3.80	3.41	3.18	9.07	6.70	5.74	5.20	17.81	12.31	10.21	9.07
14	4.60	3.74	3.34	3.11	8.86	6.51	5.56	5.03	17.14	11.78	9.73	8.62
15	4.54	3.68	3.29	3.06	8.68	6.36	5.42	4.89	16.59	11.34	9.34	8.25

df₂												
16	4.49	**3.63**	**3.24**	3.01	8.53	6.23	5.29	4.77	16.12	10.97	9.00	7.94
17	4.45	**3.59**	**3.20**	2.96	8.40	6.11	5.18	4.67	15.72	10.66	8.73	7.68
18	4.41	**3.55**	**3.16**	2.93	8.28	6.01	5.09	4.58	15.38	10.39	8.49	7.46
19	4.38	**3.52**	**3.13**	2.90	8.18	5.93	5.01	4.50	15.08	10.16	8.28	7.26
20	4.35	**3.49**	**3.10**	2.87	8.10	5.85	4.94	4.43	14.82	9.95	8.10	7.10
21	4.32	**3.47**	**3.07**	2.84	8.02	5.78	4.87	4.37	14.59	9.77	7.94	6.95
22	4.30	**3.44**	**3.05**	2.82	7.94	5.72	4.82	4.31	14.38	9.61	7.80	6.81
23	4.28	**3.42**	**3.03**	2.80	7.88	5.66	4.76	4.26	14.19	9.47	7.67	6.69
24	4.26	**3.40**	**3.01**	2.78	7.82	5.61	4.72	4.22	14.03	9.34	7.55	6.59
25	4.24	**3.38**	**2.99**	2.76	7.77	5.57	4.68	4.18	13.88	9.22	7.45	6.49
26	4.22	**3.37**	**2.98**	2.74	7.72	5.53	4.64	4.14	13.74	9.12	7.36	6.41
27	4.21	**3.35**	**2.96**	2.73	7.68	5.49	4.60	4.11	13.61	9.02	7.27	6.33
28	4.20	**3.34**	**2.95**	2.71	7.64	5.45	4.57	4.07	13.50	8.93	7.19	6.25
29	4.18	**3.33**	**2.93**	2.70	7.60	5.42	4.54	4.04	13.39	8.85	7.12	6.19
30	4.17	**3.32**	**2.92**	2.69	7.56	5.39	4.51	4.02	13.29	8.77	7.05	6.12
40	4.08	**3.23**	**2.84**	2.61	7.31	5.18	4.31	3.83	12.61	8.25	6.60	5.70
60	4.00	**3.15**	**2.76**	2.52	7.08	4.98	4.13	3.65	11.97	7.76	6.17	5.31
120	3.92	**3.07**	**2.68**	2.45	6.85	4.79	3.95	3.48	11.38	7.31	5.79	4.95
∞	3.84	**2.99**	**2.60**	2.37	6.64	4.60	3.78	3.32	10.83	6.91	5.42	4.62

Source: Adapted from **Table V of Fisher and Yates**, Statistical Tables for *Biological, Agricultural and Medical Research*, published by Longman Group **Ltd., London (previously published** by Oliver and Boyd Ltd., Edinburgh). Used by permission of the authors and publishers.

11

Chi-square

How to Test for a Relationship Between Two Nominal Variables or "Goodness of Fit" in One Nominal Variable

We have been looking in our hypothesis testing units at situations in which we have two or more sets of scores or means to compare, either (1) one-sample z and t tests, in which a sample mean is compared to a population mean, (2) two-sample z and t tests, in which one sample mean is compared to a second sample mean, or (3) analysis of variance, in which more than two samples of scores are compared. We don't always have means or scores to compare. In an earlier unit, we reviewed how measurement occurs at several different levels: nominal, ordinal, or interval/ratio.

REMEMBER:

In NOMINAL level measurement, we have our lowest level of measurement. There are categories for each variable. These categories have no mathematical relationship to one another, e.g., categories of the variable gender (male, female); or categories of the variable religion (Protestant, Catholic, Jewish, other).

In ORDINAL level measurement, the categories of a variable can be ranked (e.g., from lowest to highest).

In INTERVAL and RATIO levels of measurement, the difference between each unit of measurement on the scale a variable is measured with is constant (e.g., test scores). We can add, subtract, multiply and divide.

Means are generally obtained when our measurement is at an interval or ratio level. Analysis of variance can be calculated when a variable is measured at an ordinal, interval, or ratio level.

But when we have *only nominal* level data, we usually do not calculate means and we do not have scores. The *chi-square* test is a technique for seeing if there is a relationship between two or among more than two nominal-level variables. If there is no relationship then the variables are *independent* of each other. If the variables are independent, then the category a case is placed into on one variable has no effect on the category a case is placed into on the other variable(s). If there is a relationship (i.e., if the variables are *dependent*) then the category a case is placed into on one variable does have an effect on the category a case is placed into on the other variable(s). (Another way of saying there is a relationship is to say that one variable is contingent upon another. That is why tables in chi-square problems are sometimes called *contingency* tables in text-books.)

For simplicity, we will examine a situation in which there are only two variables. We need to have at least two variables to use chi-square in this manner. There is also a one-variable use of the chi-square test reviewed in a latter section of this chapter. The following is a typical chi-square problem.

Illustrative Problem

A. The Table

The following table contains hypothetical data which shows the relationship between gender and pet ownership.

		GENDER		
		Male	Female	Total
PET OWNERSHIP	Yes	a) 15	b) 30	45
	No	c) 20	d) 10	30
	Total	35	40	75

If you look at the table, you can see two variables. Gender, the first variable, is at the top of the table and is divided into two categories which form two columns: male and female. Pet ownership, the second variable, is at the side of the table and is divided into two categories which form two rows: yes, a pet is owned; and no, a pet is not owned. Categories should not overlap. If you are constructing the table, you should be able to place every case clearly into only one category of each variable. Otherwise, the chi-square test is not an appropriate technique.

You can also see the second thing we must have in order to do a chi-square test. We must have frequency counts in the middle of the table for different *cells* of the table. By looking at the above table, you can see that 15 males said that they owned a pet contrasted with 30 females. Twenty males said that they did not own a pet contrasted with 10 females. Those four numbers are the frequencies in the cells of this particular table. This is called a 2 by 2 table. This means that there are two categories for the row variable (pet ownership) and two categories for the column variable (gender). In a 2 by 2 table, there will always be four cells, and therefore four frequencies. (In a larger table, there will be more cells and more frequencies. If you have a 2 by 3 table with two row and three column categories, you will have six cells. You can have any number of rows and columns if there are enough people in the study to have reasonably sized frequencies in most cells.)

The other new word you should learn is the *marginals* of the table. These are the sub-totals for each row and column. In this table, the marginals are the numbers 35, 40, 45, and 30. There is also a grand total in this table of 75 (total number of people in the study). The grand total will equal the sum of the marginals or sub-totals of the rows added together (e.g., $45 + 30 = 75$, or pet owners plus non-owners equal grand total.) The grand total will also equal the sum of the marginals or sub-totals of the columns added together (e.g., $35 + 40 = 75$, or males plus females equal grand total).

B. The procedure

To set up a test of a hypothesis, using chi-square, go through the same set of steps we used previously with other hypothesis-testing techniques:

1) Construct a *null hypothesis*. The null hypothesis should state that the variables are not related or are independent of one another (i.e., the chi-square is not larger than would be expected by chance).

2) Construct an *alternative or research hypothesis*. The alternative or research hypothesis should state that the variables are related or are dependent (i.e., the chi-square is larger than would be expected by chance).

3) Choose a *test statistic*. Select *chi-square* because we have nominal level measurement of our variables and we have frequency counts for each category of the variables.

4) Choose a *significance level* (either 0.05, 0.01, or 0.001) Remember that the larger the significance level (0.05 is the largest), the greater your chance of rejecting your null hypothesis. With chi-square tests, we do not have to worry about one-tailed versus two-tailed tests. Chi-square tests are always one-tailed. We are always asking if the chi-square is *larger* than would be expected by chance. Let us select 0.05 for this example.

REMEMBER:

A *one-tailed test* has a *directional* research hypothesis. Words like "higher," "lower," "greater," "fewer" are used in the research hypothesis. In a one-tailed test, the proportion of area of the distribution that makes up the size of the significance level is placed at either end (tail) of the curve but not both ends.

A *two-tailed test* has a *nondirectional* research hypothesis. Words like "difference" or "relationship" are used in the research hypothesis. In a two-tailed test, the proportion of area of the distribution that makes up the size of the significance level is divided in half. Half is placed in each end (tail) of the curve.

---

5) Select a *sampling distribution.* When using a chi-square test, the sampling distribution will always be the chi-square distribution. Chi-square distributions vary, depending upon the number of rows and columns in the contingency table. Therefore you must specify the *degrees of freedom* (df) of the chi-square distribution you are using. The degrees of freedom are obtained by multiplying $(r - 1) \times (c - 1)$. In other words, count the number of rows in your table. Only categories of the row variable count as rows. Marginals don't count. In our 2 by 2 table, there are two rows. Then count the number of columns in your table. Again, only categories of the column variable count as columns. Marginals don't count. In our 2 by 2 table, we have two columns. We subtract 1 from the number of rows, then subtract 1 from the number of columns and then multiply thus:

$$(2 - 1) \times (2 - 1) = (1) \times (1) = 1$$

Therefore, *in any 2 by 2 table, we have 1 degree of freedom.* (In any 2 by 3 table, we always have 2 degrees of freedom.)

6) Select a *critical value.* As with our other statistics for hypothesis testing, we select a critical value with which we will compare our calculated value. As with other critical values, if our calculated value is equal to or greater than our critical value we reject our null hypothesis and accept our research hypothesis. To obtain a critical value, look at a chi-square distribution table (see end of chapter). Look up your critical value according to the degrees of freedom in your distribution and the probability level you have selected. With 1 degree of freedom (which will always be the degrees of freedom with a 2 by 2 table) at the 0.05 significance level, for example, the critical

value is 3.841. (With 1 degree of freedom at the 0.01 level, the critical value is always 6.635.)

7) We now *carry out our test*. The main purpose of the chi-square test is to calculate if the frequencies you see in the cells of the table are substantially different from the frequencies that you would expect to see in the table based upon probability. If they are, then something other than probability is going on, that is, the variables are related, or dependent. If there is very little difference between the frequencies you see in the cells of the table and the frequencies you would expect to find there if you were taking nothing but probability into consideration, then there is no relationship between those variables. In other words, they are independent.

To carry out a chi-square test, do the following:

a) Make a series of 6 column headings that look like this:

Cell	f_o	f_e	$f_o - f_e$	$(f_o - f_e)^2$	$(f_o - f_e)^2/f_e$

b) The first column is for listing each cell of the table. In our table, the cells have been named a, b, c, and d. List these in the column headed "cell." (If you have more than four cells, use further letters of the alphabet to label your cells—e, f, etc.)

Cell	f_o	f_e	$f_o - f_e$	$(f_o - f_e)^2$	$(f_o - f_e)^2/f_e$
a					
b					
c					
d					

c) The second column is for listing the *observed frequencies* corresponding to each cell of the table. These are taken directly from the table. For our table, they should look like this:

Cell	f_o	f_e	$f_o - f_e$	$(f_o - f_e)^2$	$(f_o - f_e)^2/f_e$
a	15				
b	30				
c	20				
d	10				

d) The third column is for listing *expected frequencies*. To calculate each of these, multiply the row marginal and column marginal corresponding to each cell and divide by the grand total. For cell a, we go across the row to the marginal 45 and down the column to the marginal 35. $(45) \times (35)/75 = 21$. Calculate this for each cell. For cell b, we go across the row to the marginal 45 and down the column to the marginal 40. $(45) \times (40)/75 = 24$. For cell c, we go across the row to the marginal 30 and down the column to the marginal 35. $(30) \times (35)/75 = 14$. For cell d, we go across the row to the marginal 30 and down the column to the marginal 40. $(30) \times (40)/75 = 16$. Our columns should now look like this:

Cell	f_o	f_e	$f_o - f_e$	$(f_o - f_e)^2$	$(f_o - f_e)^2/f_e$
a	15	21			
b	30	24			
c	20	14			
d	10	16			

e) The fourth column is for listing the difference between the observed frequency and the expected frequency for each cell. For each cell, subtract the number in the f_e column from the number in the f_o column. For example: $15 - 21 = -6$ in cell a. $30 - 24 = 6$ in cell b. Don't worry about minus signs. Our columns should now look like this:

Cell	f_o	f_e	$f_o - f_e$	$(f_o - f_e)^2$	$(f_o - f_e)^2/f_e$
a	15	21	– 6		
b	30	24	6		
c	20	14	6		
d	10	16	– 6		

f) In the fifth column, $(f_o - f_e)^2$, square each number from the fourth column. For example, $(-6)^2 = 36$. Numbers in this column will not always have identical values as they do for this table. For our table, our columns should now look like this:

Cell	f_o	f_e	$f_o - f_e$	$(f_o - f_e)^2$	$(f_o - f_e)^2/f_e$
a	15	21	– 6	36	
b	30	24	6	36	
c	20	14	6	36	
d	10	16	– 6	36	

g) In the last column, divide each number in the fifth column by the number in the third, f_e, column for that cell. For cell a, 36/21 = 1.714. For cell b, 36/24 = 1.500. For cell c, 36/14 = 2.571. For cell d, 36/16 = 2.250. For our table, the completed columns look like this:

Cell	f_o	f_e	$f_o - f_e$	$(f_o - f_e)^2$	$(f_o - f_e)^2/f_e$
a	15	21	– 6	36	1.714
b	30	24	6	36	1.500
c	20	14	6	36	2.571
d	10	16	– 6	36	2.250
					$\sum = 8.035$

h) Sum the values in the last column. This sum is your calculated chi-square value. In this case, 1.714 + 1.500 + 2.571 + 2.250 = 8.035.

These steps are based upon the following formula for calculating chi-square values:

$$\text{Chi-square} = \text{Sum of: } \frac{\text{Squared: (observed freq. – expected freq.)}}{\text{expected frequencies}}$$

or, in symbolic form,

$$\chi^2 = \sum \frac{(f_o - f_e)^2}{f_e} = 8.035$$

8) We compare 8.035 (our calculated value) to the critical value (the distribution table value) of 3.841, since we have selected a probability level of 0.05. Since our calculated value is higher than the critical value, the decision we make is to reject our null hypothesis and accept our research hypothesis. We conclude that there is a relationship between gender and pet ownership. If we look at the figures in the cells of our table, we can see that females are more likely to have pets.

One-Variable Use of Chi-Square

In this chapter's first example of the use of chi-square there were two variables. Suppose we only have one variable. Chi-square can also be used when there is only *one nominal variable* to test "*goodness of fit*" (i.e., to test if a significant difference exists between observed frequencies in a table and expected frequencies).

In the following table, there are frequencies for one variable: gender.

	Male	Female	Total
Number of persons	20	40	60

We would set up our hypothesis testing steps in the usual manner.

1) The null hypothesis is that there is an equal distribution of both males and females.

2) The research hypothesis is that the distribution of males and females is not equal.

3) The test statistics is the chi-square one-variable test.

4) A significance level of 0.05, 0.01 or 0.001 would be chosen. Let us choose 0.01 for this example.

5) The sampling distribution is the chi-square distribution. There is a difference in the way degrees of freedom are calculated in the chi-square one-variable test. The degrees of freedom are now the number of categories minus one. For example, in the above illustration, there are two categories: (1) male, (2) female. Therefore, the degrees of freedom equals one.

6) The critical value, with which we will be comparing our calculated value, is looked up on the chi-square distribution table according to the significance level selected and the degrees of freedom. At the .01 level, for example, the critical value is 6.635 at 1 degree of freedom (df).

7) When we carry out our test, expected frequencies are calculated by taking the total and dividing by the number of categories. In the above illustration, the expected frequencies for each cell would be $60/2 = 30$. *The expected frequencies for all categories are equal.*

Using the same chi-square formula as in the two-variable case, we would set up a series of six columns, like before. They would now look like this:

Cell	f_o	f_e	$f_o - f_e$	$(f_o - f_e)^2$	$(f_o - f_e)^2/f_e$
a	20	30	− 10	100	3.333
b	40	30	10	100	3.333
					$\sum = 6.666$

When we sum the values in the last column, we get $\sum(f_o - f_e)^2/f_e$. In this case, $\chi^2 = \sum(f_o - f_e)^2/f_e = 3.333 + 3.333 = 6.666$

8) We compare 6.666 (our calculated value) to the critical value (the distribution table value of 6.635 since we selected a probability level of 0.01). Since our calculated value is higher than the critical value, the decision we make is to reject our null hypothesis and accept our research hypothesis. We conclude that the distribution of males and females is not equal.

Review Questions

1. When should you use a chi-square test?
2. What does it mean if two variables are independent?
3. Are chi-square tests always one-tailed or two-tailed?
4. How do you compute degrees of freedom for a two-variable chi-square test? For a one-variable chi-square test?
5. What are cells?
6. What are marginals?
7. How do you compute expected frequencies?
8. What goes in the last column of the computations?

Review Exercises

Exercise 1. A scientist is interested in the relationship between gender and interest in ecological issues. The following data are collected:

		Interest in Ecology		
		High	Low	Total
Gender	Male	40	60	100
	Female	60	40	100
	Total	100	100	200

Test the hypothesis that there is a relationship between the variables.

1) The null hypothesis: The variables are not dependent.
2) The research hypothesis: The variables are dependent.
3) The test statistic: the chi-square test
4) The significance level: 0.01
5) The sampling distribution: the chi-square distribution with 1 degree of freedom ($df = (r - 1) \times (c - 1) = (1) \times (1) = 1$)
6) The critical value: 6.635
7) To carry out our test:
   a) We make 6 columns like the following:

Cell	f_o	f_e	$f_o - f_e$	$(f_o - f_e)^2$	$(f_o - f_e)^2/f_e$

b–c) The first column lists each cell of the table. The second column lists the observed frequencies from the table:

Cell	f_o	f_e	$f_o - f_e$	$(f_o - f_e)^2$	$(f_o - f_e)^2/f_e$
a	40				
b	60				
c	60				
d	40				

d) For the f_e column, we compute expected frequencies:

cell a: $f_e = (100) \times (100)/200 = 50$
cell b: $f_e = (100) \times (100)/200 = 50$
cell c: $f_e = (100) \times (100)/200 = 50$
cell d: $f_e = (100) \times (100)/200 = 50$

e) In the $f_o - f_e$ column, we compute the differences between the observed frequencies and expected frequencies:

Cell	f_o	f_e	$f_o - f_e$	$(f_o - f_e)^2$	$(f_o - f_e)^2/f_e$
a	40	50	– 10		
b	60	50	10		
c	60	50	10		
d	40	50	– 10		

f) We square each number from the $f_o - f_e$ column for the $(f_o - f_e)^2$ column:

Cell	f_o	f_e	$f_o - f_e$	$(f_o - f_e)^2$	$(f_o - f_e)^2/f_e$
a	40	50	– 10	100	
b	60	50	10	100	
c	60	50	10	100	
d	40	50	– 10	100	

g) We divide each result from step 7f by the expected frequency for that cell:

Cell	f_o	f_e	$f_o - f_e$	$(f_o - f_e)^2$	$(f_o - f_e)^2/f_e$
a	40	50	– 10	100	2
b	60	50	10	100	2
c	60	50	10	100	2
d	40	50	– 10	100	2
					$\sum = 8$

h) The sum of the values in the final column = 8.

8) The calculated value of 8 is higher than the critical value of 6.635. Therefore we reject the null hypothesis and accept the research hypothesis. We conclude that there is a relationship between gender and interest in ecological issues.

Exercise 2. A pollster is interested in testing the hypothesis that region of the country is related to political preference. A sample is taken of 100 voters on both coasts and they are classified by region and political preference. The following table results:

		Region		
		Eastern	Western	Total
Political Party	Democrat	30	15	45
	Republican	15	30	45
	No preference	5	5	10
	Total	50	50	100

Test the hypothesis that there is a relationship between the two variables.

1) The null hypothesis: The variables are not dependent.
2) The research hypothesis: The variables are dependent.
3) The test statistic: the chi-square test
4) The significance level: 0.05
5) The sampling distribution: the chi-square distribution with 2 degrees of freedom ($df = (r-1) \times (c-1) = (2) \times (1) = 2$)
6) The critical value: 5.991
7) To carry out our test:

a) We make 6 columns like the following:

Cell	f_o	f_e	$f_o - f_e$	$(f_o - f_e)^2$	$(f_o - f_e)^2/f_e$

b–c) The first column lists each cell of the table. The second column lists the observed frequencies from the table:

Cell	f_o	f_e	$f_o - f_e$	$(f_o - f_e)^2$	$(f_o - f_e)^2/f_e$
a	30				
b	15				
c	15				
d	30				
e	5				
f	5				

d) For the f_e column, we compute expected frequencies:

cell a: $f_e = (45) \times (50)/100 = 22.5$
cell b: $f_e = (45) \times (50)/100 = 22.5$
cell c: $f_e = (45) \times (50)/100 = 22.5$
cell d: $f_e = (45) \times (50)/100 = 22.5$
cell e: $f_e = (10) \times (50)/100 = 5$
cell f: $f_e = (10) \times (50)/100 = 5$

e) In the $f_o - f_e$ column, we compute the differences between the observed frequencies and expected frequencies:

Cell	f_o	f_e	$f_o - f_e$	$(f_o - f_e)^2$	$(f_o - f_e)^2/f_e$
a	30	22.5	7.5		
b	15	22.5	– 7.5		
c	15	22.5	– 7.5		
d	30	22.5	7.5		
e	5	5	0		
f	5	5	0		

f) We square each number from the $f_o - f_e$ column for the $(f_o - f_e)^2$ column:

Cell	f_o	f_e	$f_o - f_e$	$(f_o - f_e)^2$	$(f_o - f_e)^2/f_e$
a	30	22.5	7.5	56.25	
b	15	22.5	– 7.5	56.25	
c	15	22.5	– 7.5	56.25	
d	30	22.5	7.5	56.25	
e	5	5	0	0	
f	5	5	0	0	

g) We divide each result from step 7f by the expected frequency for that cell:

Cell	f_o	f_e	$f_o - f_e$	$(f_o - f_e)^2$	$(f_o - f_e)^2/f_e$
a	30	22.5	7.5	56.25	2.5
b	15	22.5	– 7.5	56.25	2.5
c	15	22.5	– 7.5	56.25	2.5
d	30	22.5	7.5	56.25	2.5
e	5	5	0	0	0
f	5	5	0	0	0
					$\sum = 10.0$

h) The sum of the values in the final column = 10.0.

8) The calculated value of 10.0 is higher than the critical value of 5.991. Therefore we reject the null hypothesis and

accept the research hypothesis. We conclude that there is a relationship between region and political preference.

Exercise 3. In a marketing test, 100 consumers were given taste tests of what was actually the same cereal in three different types of packaging: brightly colored, pastel colored or black and white boxes. Their preferences for the three supposedly different cereals were as follows:

	Preferred cereal			
	Brightly colored	Pastel colored	Black and white	Total
Number of consumers	20	35	45	100

Should the company alter its packaging, which is now brightly colored?

1) The null hypothesis: There is an equal preference for each type of packaging.

2) The research hypothesis: The preference for type of packaging is not equally distributed.

3) The test statistic: the chi-square one-variable test

4) The significance level: 0.05

5) The sampling distribution: the chi-square distribution with 2 degrees of freedom ($df = \text{number of categories} - 1 = 3 - 1 = 2$)

6) The critical value: 5.991

7) To carry out our test, we make 6 columns like the following:

Cell	f_o	f_e	$f_o - f_e$	$(f_o - f_e)^2$	$(f_o - f_e)^2/f_e$
a	20	33.3	– 13.3	176.89	5.312
b	35	33.3	1.7	2.89	.087
c	45	33.3	11.7	136.89	4.111
					$\sum = 9.510$

Expected frequencies for all categories are equal and are obtained by dividing the total number of consumers by the number of categories. 100/3 = 33.3.

The sum of the values in the last column = 5.312 + .087 + 4.111 = 9.510.

8) The calculated value of 9.510 is higher than the critical value of 5.991. Therefore we reject the null hypothesis and accept the research hypothesis. We conclude that there is a preference for type of packaging.

Table 11.1 Chi-square Distribution Table

Degrees of freedom (df)	Probability level		
	0.05	0.01	0.001
1	3.841	6.635	10.827
2	5.991	9.210	13.815
3	7.815	11.341	16.268
4	9.488	13.277	18.465
5	11.070	15.086	20.517
6	12.592	16.812	22.457
7	14.067	18.475	24.322
8	15.507	20.090	26.125
9	16.919	21.666	27.877
10	18.307	23.209	29.588

Source: Adapted from Table IV of Fisher and Yates, *Statistical Tables for Biological. Agricultural and Medical Research*, published by Longman Group Ltd., London (previously published by Oliver and Boyd Ltd., Edinburgh). Used by permission of the authors and publishers.

12

Correlation

How to Compute the Degree of Relationship Between Two Variables Using Pearson Correlation and Spearman Correlation

In this chapter, we review how to compute the *degree of association, or relationship*, between two variables which are either at the *ordinal or interval/ratio level of measurement*.

REMEMBER:

In ORDINAL level measurement, the categories of a variable can be *ranked* (i.e., from lowest to highest)

In INTERVAL and RATIO levels of measurement, the *difference between each unit* of measurement on the scale that a variable is measured with is *constant* (e.g., test scores).

There are several other techniques for computing correlations. But Pearson and Spearman correlation coefficients are the most common. They are both methods for computing *linear relationships*, that is, relationships which would form a straight line when graphed.

Calculating the degree of correlation between two variables

results in a number somewhere between 0.00 and 1.00. 0.00 tells us that there is no relationship between the variables. 1.00 tells us that there is a perfect relationship between the variables. Usually correlations fall somewhere in between. But the closer the correlation is to 1.00, the stronger the relationship between the variables. A number representing the degree of correlation can have a positive sign (*positive relationship*) or a negative sign (*negative relationship*).

A positive linear correlation means that those who score high on the first variable usually score high on the second variable also. It also means that those who score low on the first variable usually score low on the second variable. A negative linear correlation means that those who score high on the first variable usually score low on the second variable, and that those who score low on the first variable usually score high on the second variable. There may also be a very strong relationship between variables, but it is inverse, or negative. To calculate a correlation, you need to have scores or ranks on two variables for each person or object.

Spearman Correlation Coefficient for Ordinal Data

If you have two variables that are *rank ordered*, you can use the *Spearman coefficient.**

Let's suppose we have ranks for 10 students on leadership and self-esteem and we want to know if there is a relationship between leadership ability and self-esteem rank.

*If you have one variable that is in ranks and one that is in interval or ratio form, you must first rank-order the interval or ratio scores to use the Spearman coefficient. If you have two scores that are the same, give each the average of the two ranks they would ordinarily receive. For example, let's suppose two students would ordinarily occupy both the second and third ranks with a tied score. Give them each the rank of 2.5.

Student	Leadership Rank (variable 1)	Self-esteem Rank (variable 2)
A	1	4
B	2	2
C	3	1
D	4	3
E	5	8
F	6	10
G	7	9
H	8	5
I	9	7
J	10	6

To compute the degree of relationship between the two rank-ordered variables, do the following:

1) Make a fourth column labeled "D" or "differences." For each pair of scores, subtract the rank on the second variable from the rank on the first variable. In this case, for student A, $1 - 4 = -3$; for student B, $2 - 2 = 0$; for student C, $3 - 1 = 2$, etc. Place the result of each subtraction in the D column. If you add up the numbers in the D column, they should sum to zero. Your columns should now look like this:

Student	Leadership Rank (variable 1)	Self-esteem Rank (variable 2)	D
A	1	4	−3
B	2	2	0
C	3	1	2
D	4	3	1
E	5	8	−3
F	6	10	−4
G	7	9	−2
H	8	5	3
I	9	7	2
J	10	6	4
			$\sum = 0$

2) Make a fifth column labeled "D^2" or "squared differences. "Square each number in the D column and place the result in the D^2 column. In this case $(-3)^2 = 9$; $(0)^2 = 0$; $(2)^2 = 4$, etc. Notice that the negative signs all disappear. Your columns should now look like this:

Student	Leadership Rank (variable 1)	Self-esteem Rank (variable 2)	D	D^2
A	1	4	−3	9
B	2	2	0	0
C	3	1	2	4
D	4	3	1	1
E	5	8	−3	9
F	6	10	−4	16
G	7	9	−2	4
H	8	5	3	9
I	9	7	2	4
J	10	6	4	16
			$\sum = 0$	$\sum = 72$

3) Add up the numbers in the D^2 column. In this case, they sum to 72.

4) Multiply the result of step 3 by the number 6. The number 6 is a constant here. In this case, $(72) \times (6) = 432$.

5) Square the number of pairs of scores. In this case, $(10) \times (10) = 100$.

6) Subtract the number 1 from the result of step 5. In this example, $100 - 1 = 99$.

7) Multiply the result of step 6 by the number of pairs of scores. In this case, $(99) \times (10) = 990$.

8) Divide the result of step 4 by the result of step 7. In this case, $432/990 = .44$.

9) Subtract the result of step 8 from the number 1. In this case, $1 - .44 = .56$. This is your correlation coefficient.

The above instructions are based upon the following formula:

$$\text{Spearman corr. coefficient} = 1 - \frac{6\ (\text{sum of squared differences})}{\text{Number of pairs of scores}\left(\text{Squared number of pairs of scores} - 1\right)}$$

or, in symbolic form,

$$r_s = 1 - \frac{6\sum D^2}{N(N^2 - 1)} = 1 - \frac{6\ (72)}{10(100 - 1)}$$

r is the symbol for correlation
r_s is the symbol for Spearman correlation coefficient
D is the symbol for differences
N is the symbol for number of *pairs* of scores here, rather than number of scores

Pearson Correlation Coefficient for Interval/Ratio Data

If you have two variables that are at the *interval or ratio level of measurement*, you can use the *Pearson coefficient* to find the degree of association between the two variables.

Let's suppose we have scores for 10 students on creativity and problem solving ability:

Student	Creativity	Problem Solving
A	11	12
B	9	14
C	4	5
D	9	7
E	3	2
F	1	3
G	6	4
H	8	9
I	13	10
J	15	11

To compute the degree of relationship between two interval or ratio variables, do the following:

1) Make a set of columns which look like this:

Student	Creativity (X)	Problem Solving (Y)	X^2	Y^2	XY

You already have the information which fills in the first three columns. In the first column is listed the numbers, letters or names corresponding to each student, object or group (in this case, students) in your study. In the second column is listed the scores on one of the variables (e.g., creativity, now referred to as the *X variable*). In the third column is listed the scores for each on the other variable (e.g., problem-solving ability, now referred to as the *Y variable*).

2) Add up the scores in the X column. In this case they sum to 79.

3) Add up the scores in the Y column. In this case they sum to 77.

4) Square each score on the X variable and enter the squares in the X^2 column. In this case, your columns should now look like this:

Student	X	Y	X^2	Y^2	XY
A	11	12	121		
B	9	14	81		
C	4	5	16		
D	9	7	81		
E	3	2	9		
F	1	3	1		
G	6	4	36		
H	8	9	64		
I	13	10	169		
J	15	11	225		
	$\sum = 79$	$\sum = 77$	$\sum = 803$		

5) Add up the scores in the X^2 column. In this case, they sum to 803.

6) Square each score on the Y variable and enter the squares in the Y^2 column. In this case, your columns should now look like this.

Student	X	Y	X^2	Y^2	XY
A	11	12	121	144	
B	9	14	81	196	
C	4	5	16	25	
D	9	7	81	49	
E	3	2	9	4	
F	1	3	1	9	
G	6	4	36	16	
H	8	9	64	81	
I	13	10	169	100	
J	15	11	225	121	
	$\sum = 79$	$\sum = 77$	$\sum = 803$	$\sum = 745$	

7) Add up the scores in the Y^2 column. In this case, they sum to 745.

8) Multiply each score on the X variable by its corresponding score on the Y variable and enter the results in the XY column. For our data, the completed columns should now look like this:

Student	X	Y	X^2	Y^2	XY
A	11	12	121	144	132
B	9	14	81	196	126
C	4	5	16	25	20
D	9	7	81	49	63
E	3	2	9	4	6
F	1	3	1	9	3
G	6	4	36	16	24
H	8	9	64	81	72
I	13	10	169	100	130
J	15	11	225	121	165
	$\sum = 79$	$\sum = 77$	$\sum = 803$	$\sum = 745$	$\sum = 741$

9) Add up the scores in the XY column. In this case, they sum to 741.

10) Multiply the result of step 2 by the result of step 3. In this case, (79) × (77) = 6083.

11) Divide the result of step 10 by the number of pairs of scores. In this case, 6083/10 = 608.3.

12) Subtract the result of step 11 from the result of step 9. In this case, 741 – 608.3 = 132.7.

13) Square the result of step 2. In this case, (79) × (79) = 6241.

14) Divide the result of step 13 by the number of pairs of scores. In this case, 6241/10 = 624.1.

15) Subtract the result of step 14 from the result of step 5. In this case, 803 – 624.1 = 178.9.

16) Square the result of step 3. In this case, (77) × (77) = 5929.

17) Divide the result of step 16 by the number of pairs of scores. In this case, 5929/10 = 592.9.

18) Subtract the result of step 17 from the result of step 7. In this case, 745 – 592.9 = 152.1.

19) Multiply the result of step 15 by the result of step 18. In this case (178.9) × (152.1) = 27,210.69.

20) Compute the square root of the result of step 19. In this case, the square root of 27,210.69 is 164.96.

21) Divide the result of step 12 by the result of step 20. In this case, 132.7/164.96 = .80. This is your Pearson correlation coefficient.

The above instructions are based upon the following formula:

Pearson r =

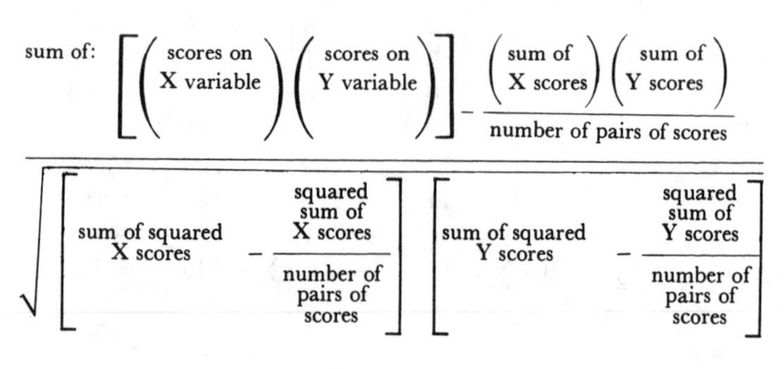

$$\frac{\text{sum of: }\left[\left(\text{scores on X variable}\right)\left(\text{scores on Y variable}\right)\right] - \dfrac{\left(\text{sum of X scores}\right)\left(\text{sum of Y scores}\right)}{\text{number of pairs of scores}}}{\sqrt{\left[\text{sum of squared X scores} - \dfrac{\text{squared sum of X scores}}{\text{number of pairs of scores}}\right]\left[\text{sum of squared Y scores} - \dfrac{\text{squared sum of Y scores}}{\text{number of pairs of scores}}\right]}}$$

or, in symbolic form,

$$r = \frac{\sum XY - \frac{(\sum X)(\sum Y)}{N}}{\sqrt{\left[\sum X^2 - \frac{(\sum X)^2}{N}\right]\left[\sum Y^2 - \frac{(\sum Y)^2}{N}\right]}} = \frac{741 - \frac{(79)(77)}{10}}{\sqrt{\left[803 - \frac{(79)^2}{10}\right]\left[745 - \frac{(77)^2}{10}\right]}}$$

$\sum X^2$ is the symbol for sum of squared scores on the X variable
$(\sum X)^2$ is the symbol for squared sum of scores on the X variable
$\sum Y^2$ is the symbol for sum of squared scores on the Y variable
$(\sum Y)^2$ is the symbol for squared sum of scores on the Y variable.

Hypothesis Testing with the Pearson Correlation Coefficient

Hypothesis testing can be used to test if a particular correlation value is significantly different from zero. The correlation may be greater than zero in the sample that was drawn, but like all other sample situations, we need to know if this was probably an artifact of the sampling process or whether the correlation value is different enough from zero so that it probably differs from zero for a reason other than chance alone. We test this type of hypothesis in a way similar to the hypothesis testing we have already seen in prior chapters. When using this test we assume that the two variables are related in a uniform manner across all values of the variables, if the variables are in fact related. It should also be noted that we are testing only for the significance of a *linear* relationship between the variables. Let us use the correlation of .80 that we obtained in the previous example:

1) Construct a *null hypothesis*. The null hypothesis here should state that there is no linear correlation between the two variables.

2) Construct an *alternative* or *research hypothesis*. The alternative or research hypothesis should state that there is a correlation between the two variables.

3) Choose a *test statistic*. We can use a form of the *t* test.

4) Choose a *significance level*. We must choose either 0.05, 0.01

or 0.001 for the tail(s). The larger your significance level (0.05 is the largest), the greater your chance of rejecting your null hypothesis. Let us select 0.01 for this example.

5) Select a *sampling distribution*. With t tests, the sampling distribution is always the t distribution. Remember that t distributions vary, depending upon the number of cases in the sample. Therefore, you must specify the *degrees of freedom* (df) of the t distribution you are using. The degrees of freedom are obtained by subtracting two from the number of pairs of scores. In this case, $df = N - 2 = 10 - 2 = 8$.

6) Select a *critical value*. We select a critical value with which we will compare our *calculated correlation value*. If our calculated value is equal to or greater than our critical value, our t value falls in a tail of the distribution. We then reject our null hypothesis and accept our research hypothesis. We find our critical value in the t distribution table at the end of Chapters 8 and 9 (Table 8.1 and Table 9.1). Look up your critical value according to (a) the degrees of freedom in your distribution, (b) the probability level you have selected, and (c) whether or not your test is one-tailed or two-tailed. With 8 degrees of freedom, at the 0.01 level of probability and a one-tailed test, for example, the critical value is 2.896.

7) We now *carry out our test*. The main purpose of this is to test if the calculated correlation value comes from a distribution of values that is different from the distribution of correlation values that would be obtained by repeated sampling from a population in which the correlation between the variables was zero.

To carry out the test, do the following:

a) Subtract 2 from the number of pairs of scores. In this case, $10 - 2 = 8$.

b) Square the calculated correlation value. In this case, $(.80)^2 = .64$.

c) Subtract the result of step 7b from the number 1. One is a constant here. In this case, $1 - .64 = .36$.

d) Divide the result of step 7a by the result of step 7c. In this case, $8/.36 = 22.22$.

e) Compute the square root of the result of step 7d. In this case, $\sqrt{22.22} = 4.71$.

f) Multiply the calculated correlation value by the result of step 7e. This is your calculated t value. In this case, $(.80) \times (4.71) = 3.768$.

These steps are based upon the following formula:

$$t = \begin{matrix}\text{calculated} \\ \text{correlation} \\ \text{value}\end{matrix} \sqrt{\frac{\text{number of pairs of scores} - 2}{1 - \text{squared calculated correlation value}}}$$

or, in symbolic form,

$$t = r\sqrt{\frac{n-2}{1-r^2}}$$

8) We compare 3.768 (our calculated value) to the critical value of 2.896. Since our calculated value is higher than the critical value, the decision we make is to reject our null hypothesis and accept our research hypothesis. We conclude that the correlation between creativity and problem solving is significantly greater than zero.

Review Questions

1. When do we use Spearman correlation and when do we use Pearson correlation?
2. What is a negative correlation?
3. What would a perfect negative correlation look like?
4. What do we do with the ranks on variables to compute a Spearman correlation?
5. What do we do with the scores on variables to compute a Pearson correlation?
6. What assumptions do we make when testing if a Pearson correlation value is significantly different from zero?

Review Exercises

Exercise 1. What is the degree of relationship between attractiveness and familiarity in the following data?

Person #	Attractiveness Rank	Familiarity Rank
1	1	2
2	2	3
3	3	4
4	4	1
5	5	8
6	6	5
7	7	6
8	8	10
9	9	7
10	10	9

1) We make a fourth column labeled "D" and subtract the rank on the second variable from the rank on the first variable for each pair of scores:

Person #	Attractiveness Rank	Familiarity Rank	D
1	1	2	−1
2	2	3	−1
3	3	4	−1
4	4	1	3
5	5	8	−3
6	6	5	1
7	7	6	1
8	8	10	−2
9	9	7	2
10	10	9	1
			$\sum = 0$

We check to see that the numbers in the D column sum to zero.

2) We make a fifth column labeled D^2 and square each number in the D column:

Person #	Attractiveness Rank	Familiarity Rank	D	D^2
1	1	2	−1	1
2	2	3	−1	1
3	3	4	−1	1
4	4	1	3	9
5	5	8	−3	9
6	6	5	1	1
7	7	6	1	1
8	8	10	−2	4
9	9	7	2	4
10	10	9	1	1
			$\sum = 0$	$\sum = 32$

3) The sum of the numbers in the D^2 column = 32.

4) The result of step 3 multiplied by the number 6 = (32) × (6) = 192.

5) The number of pairs of scores squared = $(10)^2 = 100$.

6) The result of step 5 minus 1 = 100 − 1 = 99.

7) The result of step 6 multiplied by the number of pairs of scores = (99) × (10) = 990.

8) The result of step 4 divided by the result of step 7 = 192/990 = .19.

9) The number 1 minus the result of step 8 = 1 − .19 = a Spearman correlation of .81.

Exercise 2. The following are the scores of 7 respondents on an achievement motivation scale and on a self-actualization scale. What is the degree of relationship between achievement motivation and self-actualization?

Respondent #	Achievement Motivation (X)	Self-Actualization (Y)
1	8	10
2	7	8
3	3	2
4	5	6
5	7	9
6	2	2
7	4	5

1) We make a set of columns which looks like this (We already have the information which fills in the first three columns):

Respondent #	Achievement Motivation (X)	Self-Actualization (Y)	X^2	Y^2	XY
1	8	10			
2	7	8			
3	3	2			
4	5	6			
5	7	9			
6	2	2			
7	4	5			

2) The sum of the scores in the X column = 36.
3) The sum of the scores in the Y column = 42.
4) We square each score in the X column and enter the squares in the X^2 column:

Respondent #	Achievement Motivation (X)	Self-Actualization (Y)	X^2	Y^2	XY
1	8	10	64		
2	7	8	49		
3	3	2	9		
4	5	6	25		
5	7	9	49		
6	2	2	4		
7	4	5	16		
	$\sum = 36$	$\sum = 42$	$\sum = 216$		

5) The sum of the scores in the X^2 column = 216.

6) We square each score in the Y column and enter the squares in the Y^2 column:

Respondent #	Achievement Motivation (X)	Self-Actualization (Y)	X^2	Y^2	XY
1	8	10	64	100	
2	7	8	49	64	
3	3	2	9	4	
4	5	6	25	36	
5	7	9	49	81	
6	2	2	4	4	
7	4	5	16	25	
	$\sum = 36$	$\sum = 42$	$\sum = 216$	$\sum = 314$	

7) The sum of the scores in the Y^2 column = 314.

8) We multiply each score on the X variable by its corresponding score on the Y variable and enter the results in the XY column:

Respondent #	Achievement Motivation (X)	Self-Actualization (Y)	X^2	Y^2	XY
1	8	10	64	100	80
2	7	8	49	64	56
3	3	2	9	4	6
4	5	6	25	36	30
5	7	9	49	81	63
6	2	2	4	4	4
7	4	5	16	25	20
	$\sum = 36$	$\sum = 42$	$\sum = 216$	$\sum = 314$	$\sum = 259$

9) The sum of the scores in the XY column = 259.

10) The result of step 2 multiplied by the result of step 3 = $(36) \times (42) = 1512$.

11) The result of step 10 divided by the number of pairs of scores = $1512/7 = 216$.

12) The result of step 9 minus the result of step 11 = $259 - 216 = 43$.

13) The result of step 2 squared = $(36)^2 = 1296$.

14) The result of step 13 divided by the number of pairs of scores = $1296/7 = 185.14$.

15) The result of step 5 minus the result of step $14 = 216 - 185.14 = 30.86$.

16) The result of step 3 squared $= (42)^2 = 1764$.

17) The result of step 16 divided by the number of pairs of scores $= 1764/7 = 252$.

18) The result of step 7 minus the result of step $17 = 314 - 252 = 62$.

19) The result of step 15 multiplied by the result of step $18 = (30.86) \times (62) = 1913.32$.

20) The square root of the result of step $19 = \sqrt{1913.32} = 43.74$.

21) The result of step 12 divided by the result of step $20 = 43/43.74 =$ a Pearson correlation of .98.

Exercise 3. Test whether the correlation value computed in Exercise 2 of this chapter is significantly greater than zero.

1) The null hypothesis: $r = 0$
2) The research hypothesis: $r > 0$
3) The test statistic: *t* test
4) The significance level: 0.05
5) The sampling distribution: *t* distribution with 5 degrees of freedom ($df = N - 2 = 7 - 2 = 5$)
6) The critical value: 2.015
7) To carry out our test:

a) The number of pairs of scores minus $2 = 7 - 2 = 5$.

b) The calculated correlation value squared $= (.98)^2 = .96$.

c) One minus the result of step $7b = 1 - .96 = .04$.

d) The result of step 7a divided by the result of step $7c = 5/.04 = 125$.

e) The square root of the result of step $7d = \sqrt{125} = 11.18$.

f) The calculated correlation value multiplied by the result of step $7e = (.98) \times (11.18) = 10.956$.

8) The calculated value of 10.956 is higher than the critical value of 2.015. Therefore we reject the null hypothesis. We conclude that the correlation between achievement motivation and self-actualization is significantly greater than zero.

13

Correlation and Prediction

The Regression Line and the Regression Equation

In Chapter 12, we learned technique of correlation—how to determine the degree of relationship between two variables when that relationship was linear. This chapter reviews the meaning of that correlation coefficient: How to visualize the relationship between two sets of scores by constructing a scatter diagram, and how to use a correlation value to predict a person's score on a second variable when only their score on the first variable is known.

Scatter Diagrams

In analyzing a relationship between two variables, it can be useful to plot a scatter diagram. This is a visual presentation of the relationship. To make a scatter diagram, we use a vertical and horizontal line of equal length. Usually, the *independent variable* (the variable we expect to influence the other variable) is labeled along the *horizontal* line and the *dependent variable* (the variable whose variation we expect to be influenced by the independent variable) is labeled along the *vertical* line. If it is unclear which variable is dependent upon the other, we would just label one along the vertical line and the other along the horizontal line. We then place a scale along each line, its values depending upon the variables we are working with. We place a dot on the scatter diagram for each pair of scores.

For example, let us use the scores on creativity and problem solving ability from the illustrated Pearson correlation problem in Chapter 12 (p. 153):

Student	Creativity	Problem Solving
A	11	12
B	9	14
C	4	5
D	9	7
E	3	2
F	1	3
G	6	4
H	8	9
I	13	10
J	15	11

1) We make our horizontal and vertical lines and label our scales:

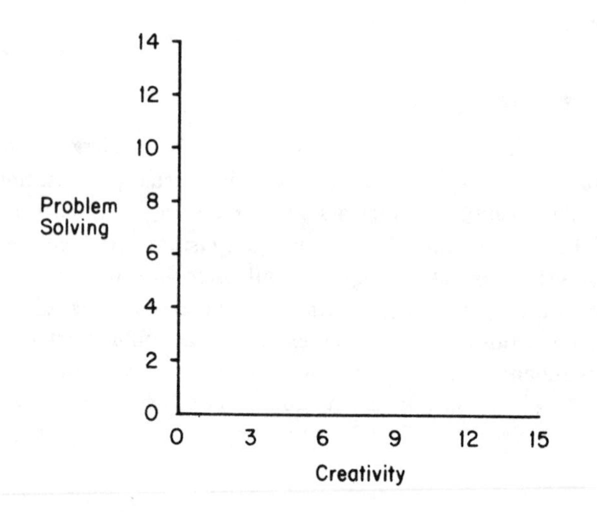

2) We place our dots on the scatter diagram:

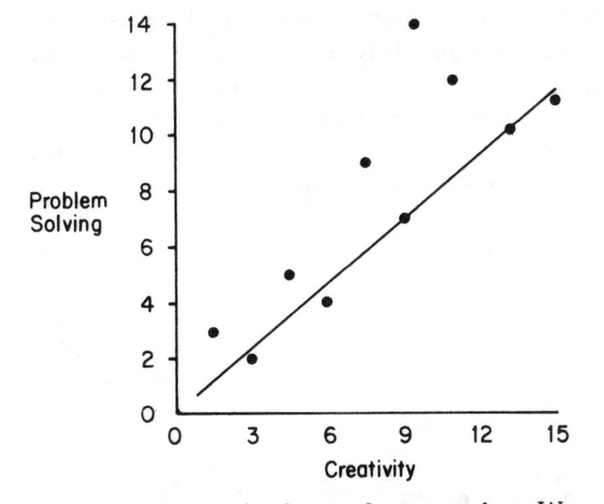

Each dot represents a pair of scores for one student. We can see that the dots fall pretty much along a straight path. This is called a *regression line*. This pattern of dots is what we would expect if our Pearson correlation value were .80. Any fairly high, positive correlation value would reflect pairs of scores which would look like the above scatter diagram when plotted. A high and negative correlation value would be reflected in scores which fall along a diagonal line, or regression line, going in the opposite direction:

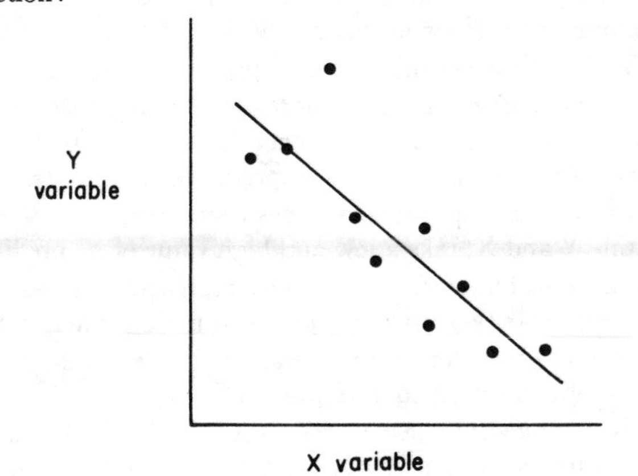

We can make an approximation to the regression line by moving a ruler around on the graph until it passes through as many dots as possible. This line can then be used to make *rough* predictions of scores on one variable from known scores on the other variable. For example, if our creativity score were 7, we could predict a problem solving score of approximately 6 by drawing a perpendicular line from the regression line to 7 on the creativity line:

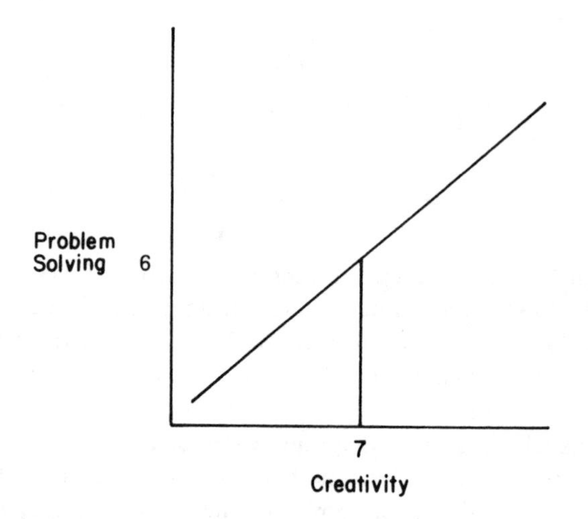

We would then look across to the vertical line to find the corresponding value for problem solving.

If we want to make more refined predictions, we can use a form of the *regression equation.* A more accurate regression line is drawn so that the squared differences (between actual scores represented by dots and the scores predicted by the regression line) are as small as possible. To construct a regression line for predicting Y and X, take a low and high value of X, predict Y values for each of those X values using a regression equation and put dots for those two sets of values on your scatter diagram. If you join these two dots with a straight line you will get the regression line for predicting Y from X.

You can follow the same procedure in reverse to get the regression line for predicting X from Y: Take a high and low value of Y, predict X values for each of those Y values using a regression equation and put dots for those two sets of values on your scatter diagram. Both regression lines will cross at the means of the X and Y variables. If your correlation were ever 1.00, both lines would overlap, rather than cross.

The following section explains how to use regression equations to predict unknown values on one variable from known values on another variable.

The Regression Equation

METHOD 1: Using means and raw score quantities from a correlation exercise.

If you have *just computed a Pearson correlation value*, the following method is the more convenient one for predicting a score on variable Y when the score on variable X is known.

Let us, again, use the calculated information from the previous chapter's illustrated Pearson correlation problem (p. 153). Let us suppose that we wish to predict the problem solving score of a student who obtains a score of 7 on the creativity measure. To predict the score we must assume that the correlation between creativity and problem solving will remain the same for new students as it has been for old students.

To predict the student's problem solving score, you must know the sums of the columns used to compute a Pearson correlation. You must also know the means of the scores on each of the variables.

To predict a score on one variable when you know the person's score on another variable, do the following:

1) Multiply the number of pairs of scores by the sum of the XY column. In this case $(10) \times (741) = 7410$.

2) Multiply the sum of the X column by the sum of the Y column. In this case $(79) \times (77) = 6083$.

3) Subtract the result of step 2 from the result of step 1. In this case, $7410 - 6083 = 1327$.

4) Multiply the number of pairs of scores by the sum of the X^2 column. In this case $(10) \times (803) = 8030$.

5) Square the sum of the X column. In this case $(79)^2 = 6241$.

6) Subtract the result of step 5 from the result of step 4. In this case, 8030 – 6241 = 1789.

7) Divide the result of step 3 by the result of step 6. In this case, 1327/1789 = .74.

8) Compute the mean of the scores on the X variable. Remember that the mean of a list of scores is obtained by dividing the sum of that list by the total number of scores. In this case, 79/10 = a mean of 7.9.

9) Subtract the result of step 8 from the score on the X variable whose corresponding score on the Y variable you are trying to find. In this case, the score on the X variable (creativity) is 7. 7 – 7.9 = – .9.

10) Multiply the result of step 7 by the result of step 9. In this case (.74) × (– .9) = – .67.

11) Compute the mean of the scores on the Y variable. In this case 77/10 = a mean of 7.7.

12) Add the result of step 10 to the result of step 11. In this case, 7.7 + (– .67) = 7.03. This is the predicted score on the Y variable (problem solving) for a score of 7 on the X variable (creativity).

The above instructions are based upon the following formula:

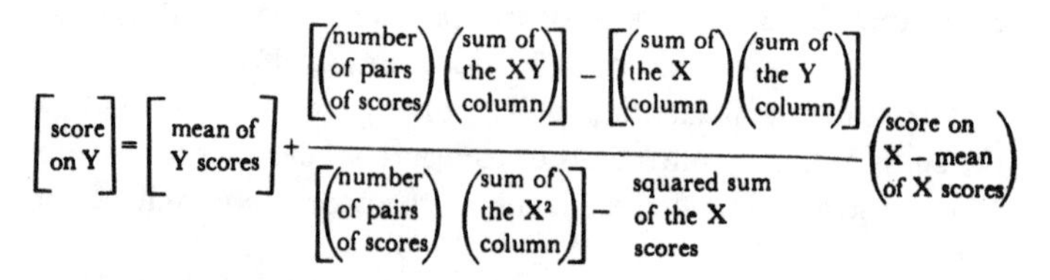

or, in symbolic form,

$$Y = \overline{Y} + \frac{N\sum XY - (\sum X)(\sum Y)}{N\sum X^2 - (\sum X)^2}(X - \overline{X})$$

$$= 7.7 + \frac{[(10)(741)] - [(79)(77)]}{[(10)(803)] - (79)^2}(7 - 7.9) = 7.03$$

If the situation is reversed and you wish to predict a value on the X variable from a known value on the Y variable, you can use the same procedure, substituting Y wherever X appears and X wherever Y appears. The formula then becomes:

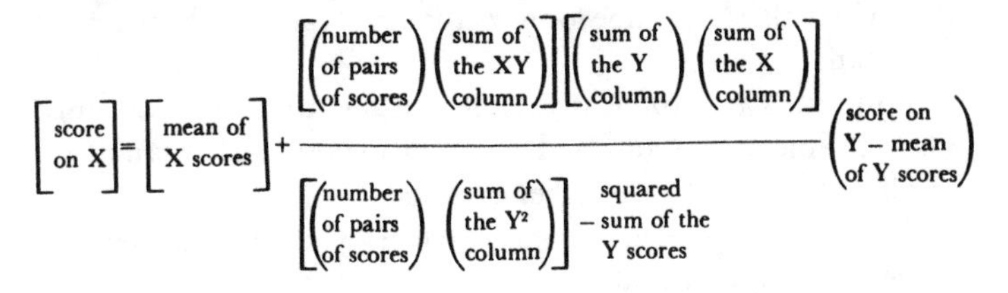

or, in symbolic form,

$$X = \overline{X} + \frac{N\sum XY - (\sum Y)(\sum X)}{N\sum Y^2 - (\sum Y)^2}(Y - \overline{Y})$$

Note that the numerator for predicting a value on the X variable will result in the same value as in the situation in which you are predicting a value on the Y variable.

METHOD 2: Using means, standard deviations and a correlation value.

The following method is the more convenient one to use in prediction if you have means and *standard deviations* of scores on each variable, as well as a correlation value.

Suppose the personnel office of a corporation has been administering an achievement test to all applicants for entry-level positions. In the past, there has been a correlation of .75 between scores on the examination and a productivity measure taken later by the applicants. If the personnel manager could predict how well new applicants will perform in their jobs, new employees could be selected accordingly.

Suppose that in this personnel office, the mean achievement score is 80 with a standard deviation of 7. The mean productivity score is 44 with a standard deviation of 5. A new applicant obtains a score of 84 on the achievement test.

To predict the applicant's score on the productivity measure, the following assumptions are made: (1) the achievement scores and productivity scores for new applicants will have the same mean and standard deviation as the scores for previous applicants; (2) the correlation between achievement and productivity will be the same for new applicants as it has been for old applicants.

To predict the applicant's score on one measure, you must know the mean and standard deviation of both variables and the applicant's score on the other measure.

To predict a score on one variable when you know the person's score on another variable, do the following:

1) Divide the standard deviation of the variable on which the score is unknown by the standard deviation of the variable on which the score is known. In this case, $5/7 = .71$.

2) Subtract the mean of the variable on which the score is known from the known score. In this case, $84 - 80 = 4$.

3) Multiply the results of steps 1 and 2. In this case, $(.71) \times (4) = 2.84$.

4) Multiply the correlation between the variables by the result of step 3. In this case, $(.75) \times (2.84) = 2.13$.

5) Add the result of step 4 to the mean of the variable on which the score is unknown. In this case, $2.13 + 44 = 46.13$. This is the mean productivity score of individuals who obtain a score of 84 on the achievement test. Particular individuals' scores will fall on both sides of the mean as in any distribution. In other words, a prediction based upon this formula (which is called a *regression equation*) is rarely exact.

The above steps are based upon the following formula for predicting a score on variable Y when the score on variable X is known:

$$\text{Score on Y} = \begin{pmatrix}\text{correlation} \\ \text{between X and} \\ \text{Y variable}\end{pmatrix} \left(\frac{\text{standard deviation of Y variable}}{\text{standard deviation of X variable}}\right) \times \begin{pmatrix}\text{Score} & & \text{Mean of} \\ \text{on} & - & \text{X} \\ \text{X} & & \text{variable}\end{pmatrix} + \begin{pmatrix}\text{mean} \\ \text{of Y} \\ \text{variable}\end{pmatrix}$$

or, in symbolic form,

$$Y = r_{xy}(s_y/s_x)(X - \overline{X}) + \overline{Y} = .75(5/7)(84 - 80) + 44 = 2.13 + 44$$
$$= 46.13$$

Y is the symbol for a person's score on the unknown variable. $\overline{Y}$ is the symbol for the mean of the unknown variable.

Obtaining the Standard Error of your Prediction

Since a prediction based upon the above formula is rarely exact, it is desirable to have a measure of the prediction's reliability. We obtain this by calculating the standard deviation of the scores on the second variable of all the individuals who obtained identical scores on the first variable. (The standard deviation is about the same no matter which score on the first variable we select, if the relationship between the variables is linear.) This is called the standard error of your prediction.

To obtain the standard error for the illustrated exercise on p. 171, do the following:

1) Square the coefficient of correlation. In this case, $(.75)^2 = .56$.

2) Subtract the result of step 1 from the number 1. The number 1 is a constant here. In this case, $1 - .56 = .44$.

3) Obtain the square root of step 2. In this case, the square root of .44 is .66.

4) Multiply the standard deviation of the variable on which the score is unknown by the results of step 3. In this case, $(5) \times (.66) = 3.3$. This is interpreted in the same way as are other standard deviations (see Chapter 5). Assuming a normal distribution for productivity scores, 68.26% of all persons who score 84 on the achievement test will score within 3.3 points of the mean productivity score predicted in this example (from 42.83 to 49.43 on the productivity measure, i.e., mean = 46.13; $46.13 - 3.3 = 42.83$; $46.13 + 3.3 = 49.43$).

The above steps are based upon the following formula for obtaining the standard error:

$$\text{standard error} = \begin{pmatrix}\text{standard deviation}\\ \text{of variable on which}\\ \text{the score is unknown}\end{pmatrix}\left(\sqrt{1 - \begin{matrix}\text{squared corr.}\\ \text{between X and}\\ \text{Y variable}\end{matrix}}\right)$$

or, in symbolic form,

$$\text{standard error} = s_y \sqrt{1 - r_{xy}^2} = 5\sqrt{1 - (.75)^2} = 3.3$$

Although we have illustrated how to predict the value on one variable with knowledge of the value on one other variable, often in research there are more than two variables involved. Often, information on *several* variables makes a prediction about the score on another variable more accurate. Correlation and regression, therefore, often involve multiple variables. They also usually involve larger numbers of cases than we have been using. However, by looking at simpler two-variable illustrations, basic principles can be illuminated.

Review Questions

1. How is a regression line drawn?
2. What assumptions do you make when using the regression equation?
3. What information must you have to use the regression equation?
4. How exact a prediction can you make when using the regression equation?
5. What does the standard error measure?

Review Exercises

Exercise 1. Make a scatter diagram representing the paired scores in Chapter 12, exercise 2.

1) We make our horizontal and vertical lines and label our scales.

2) We place our dots on the scattergram:

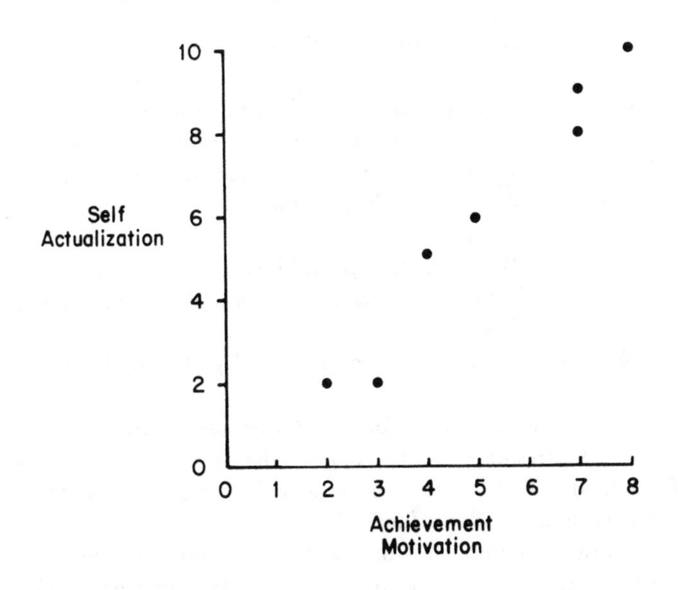

Exercise 2. Take two values of achievement motivation from Chapter 12, exercise 2, and compute predicted self-actualization scores. Use these to draw a regression line for predicting Y from X.

Score 1: Let us select a score of 2.

1) Number of pairs of scores multiplied by the sum of the XY column = (7) × (259) = 1813.

2) The sum of the X column multiplied by the sum of the Y column = (36) × (42) = 1512.

3) The result of step 1 minus the result of step 2 = 1813 – 1512 = 301.

4) The number of pairs of scores multiplied by the sum of the X^2 column = (7) × (216) = 1512.

5) The squared sum of the X column = $(36)^2$ = 1296.

6) The result of step 4 minus the result of step 5 = 216.

7) The result of step 3 divided by the result of step 6 = 301/216 = 1.39.

8) The mean of the scores on the X variable = 36/7 = 5.14.

9) The score on the X variable minus the result of step 8 = 2 – 5.14 = – 3.14.

10) The result of step 7 multiplied by the result of step 9 = (1.39) × (– 3.14) = – 4.36.

11) The mean of scores on the Y variable = 42/7 = 6.0.

12) The result of step 11 plus the result of step 10 = 6.0 + (– 4.36) = 1.64. This is your predicted score on the Y variable for a score on the X variable of 2.

Score 2: Let us select a score of 8.

1) through 8) the first 8 steps will be the same as for score 1.

9) The score on the X variable minus the result of step 8 = 8 – 5.14 = 2.86.

10) The result of step 7 multiplied by the result of step 9 = (1.39) × (2.86) = 3.98.

11) The mean of scores on the Y variable = 42/7 = 6.0.

12) The result of step 11 plus the result of step 10 = 6.0 + 3.98 = 9.98. This is your predicted score on the Y variable for a score on the X variable of 8.

We draw our regression line by placing (1) one dot where 2 on the horizontal axis and 1.64 on the vertical axis intersect, and (2) another dot where 8 on the horizontal axis and 9.98 on the vertical axis intersect. We then connect these two dots with a straight line:

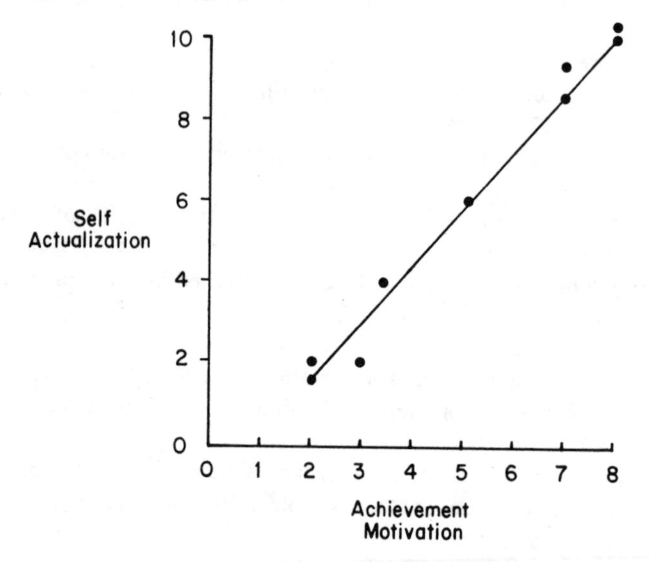

Exercise 3. In X university, the mean score of freshman students on a national exam taken the previous year is 675 with a standard deviation of 23. The mean grade point average of the freshman students is 3.2 with a standard deviation of .55. The correlation between national exam scores and grade point average is .88. A new applicant obtains a score of 640 on the national exam. Predict that applicant's grade point average.

1) The standard deviation of grade point average divided by the standard deviation of the national exam $= .55/23 = .02$.

2) The applicant's national exam score minus the mean national exam score $= 640 - 675 = -35$.

3) The result of step 1 multiplied by the result of step 2 $= (.02) \times (-35) = -.7$.

4) The correlation between the variables multiplied by the result of step 3 $= (.88) \times (-.7) = -.62$.

5) The sum of the result of step 4 and the mean grade point average $= -.62 + 3.2 =$ a mean grade point average of 2.58 for individuals who obtain a score of 640 on the national exam.

Exercise 4. Compute the standard error of the prediction in exercise 3 of this chapter.

1) The correlation squared $= (.88)^2 = .77$.

2) The number 1 minus the result of step 1 $= 1 - .77 = .23$.

3) The square root of .23 is .48.

4) The standard deviation of grade point average multiplied by the result of step 3 $= (.55) \times (.48) = .26$. Assuming a normal distribution for grade point average, 68.26% of all persons who score 640 on the national exam will score within .26 points of the mean grade point average of persons who score 640 (from 2.32 to 2.84). $2.58 - .26 = 2.32$; $2.58 + .26 = 2.84$.

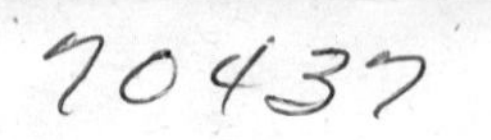
70437

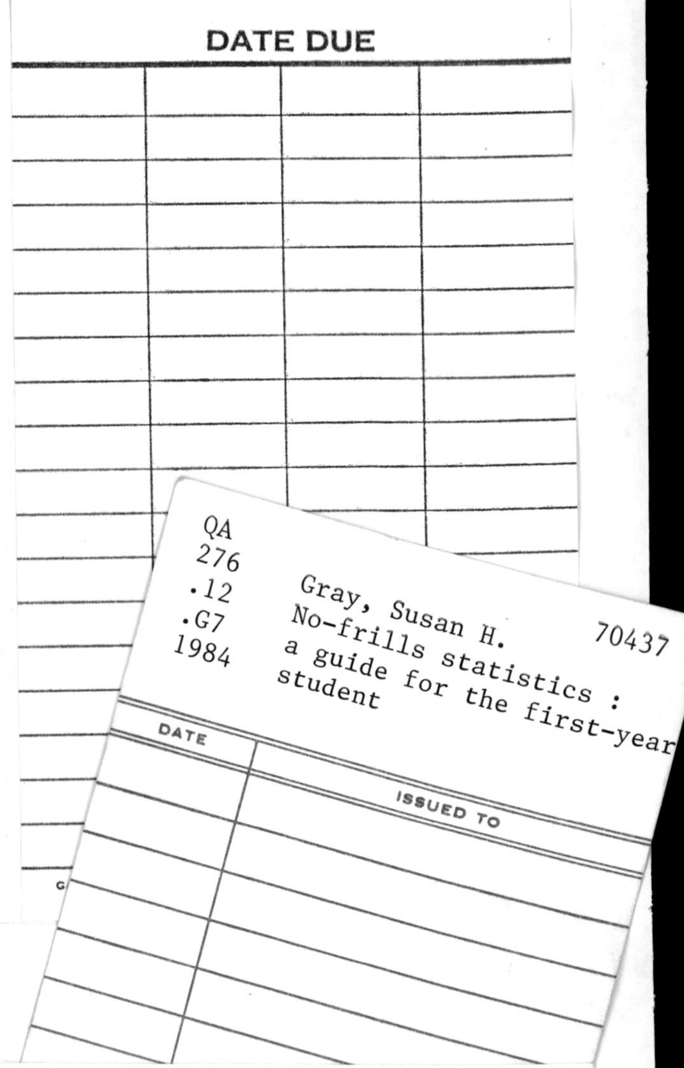
DATE DUE
QA
276
.12
.G7
1984
Gray, Susan H. 70437
No-frills statistics :
a guide for the first-year
student
DATE
ISSUED TO